BIOLOGY AS IDEOLOGY

BIOLOGY AS IDEOLOGY
THE DOCTRINE OF DNA

R . C . L E W O N T I N

HarperPerennial
A Division of HarperCollins*Publishers*

Library of Congress Cataloging-in-Publication Data

Lewontin, Richard C., 1929–
    Biology as ideology : the doctrine of DNA / by R.C. Lewontin.—1st U.S. ed.
        p. cm.
    Originally published: Concord, Ontario : Anansi, 1991, in series: The Massey lecture series.
    Includes bibliographical references.
    ISBN 0-06-097519-9 (pbk.)
    1. Biology—Philosophy.   2. Biology—Social aspects.   I. Title.
QH331.L535   1993
574'01—dc20   92–54487

06 07 08 RRD H 30 29 28 27 26 25 24 23 22 21

Contents

Preface

For a large part of the history of Western culture, the chief sources of popular consciousness about society were tradition and the Christian Church. Even social revolutionaries like the ideologues of the American Revolution appealed to divine providence for a justification of their politics. In the present century, however, Western society has become more secular and more rationalist, and the chief sources for social theory have become the professional intellectuals, the scientists, economists, political theorists, and philosophers who work largely in universities. These intellectuals are aware of the power they have to mold public consciousness, and they constantly seek ways in which they can publicize their ideas. The common pathway is to become a minor celebrity, known for some all-encompassing and usually rather simplistic "discovery" about the secret of human social and psychic existence. It's all sex or money or genes. A simple and dramatic theory that explains everything makes good press, good radio, good TV, and best-selling books. Anyone with academic authority, a halfway decent writing style, and a simple and powerful idea has easy entry to the public consciousness.

On the other hand, if one's message is that things are complicated, uncertain, and messy, that no simple rule or force will explain the past and predict the future of human existence, there are rather fewer ways to get that message across. Measured claims about the complexity of life and our ignorance of its determinants are not show biz.

Fortunately, there is a tradition, of which the Massey Lectures are an important part, of providing a public forum for a more complex and less showy world view. So, I was both flattered and delighted to be invited to give the 1990 Massey Lectures on the CBC and to turn the lectures into this book. That invitation has provided an opportunity to struggle against the view that science consists of simple objective truths and that if only we will listen to biologists we will know everything worth knowing about human existence.

The rhetoric, and especially the written rhetoric, of science is very different from ordinary forms of communication, so it was especially difficult to create a set of radio lectures that would be accessible to listeners in general. There was then the further difficulty of delivering those lectures in a lively manner to an unseen radio audience. To the extent that I have succeeded in these tasks, I owe it to the critical judgment of Jill Eisen of the CBC, who made me do it over again until it was right. Without her work and encouragement, the lectures would have been a total failure.

In turning the lectures into this book, I have resisted the temptation to lapse back into the formal rhetoric of written intellectual production, and instead have kept the radio lectures pretty much as they were. Inevitably that leads to some discursiveness and lack of clarity. I am tremendously indebted to Shaun Oakey for editing the text with a touch that was both light and unerring. He made great improvements.

For this edition of *Biology as Ideology* I have added to the original Massey Lectures an edited version of a recent essay that appeared in *The New York Review of Books*. While the framework for the essay was that of a book review, it is, in fact, a discussion of important policy consequences of the belief in the power of DNA. The rhetoric of that essay is the more studied style of written English, but it was intended for the same audience as the original Massey Lectures. I am grateful to Wendy Wolf for having urged me to add this chapter and for her help in doing so.

Finally, I am indebted to Rachel Nasca for her production of several versions of the lectures and text from tapes and illegible scrawls. Nothing would have been produced without her.

R.C. LEWONTIN
Cambridge, Massachusetts
16 July 1991

Marlboro, Vermont
27 August 1992

BIOLOGY AS IDEOLOGY

A REASONABLE SKEPTICISM

Science is a social institution about which there is a great deal of misunderstanding, even among those who are part of it. We think that science is an institution, a set of methods, a set of people, a great body of knowledge that we call scientific, is somehow apart from the forces that rule our everyday lives and that govern the structure of our society. We think science is objective. Science has brought us all kinds of good things. It has tremendously increased the production of food. It has increased our life expectancy from a mere 45 years at the beginning of the last century to over 70 in rich places like North America. It has put people on the moon and made it possible to sit at home and watch the world go by.

At the same time, science, like other productive activities, like the state, the family, sport, is a social institution completely integrated into and influenced by the structure of all our other social institutions. The problems that science deals with, the ideas that it uses in investigating those problems, even the so-called scientific results that come out of scientific investigation, are all deeply influenced by predispositions that derive from the society in which we live. Scientists do not begin life as scientists, after all, but as social beings immersed in a family, a state, a productive structure, and they view nature through a lens that has been molded by their social experience.

Above that personal level of perception, science is molded by society because it is a human productive activity that takes time and money, and so is guided by and directed by those forces in the world that have control over money and time. Science uses commodities and is part of the process of commodity production. Science uses money. People earn their living by science, and as a consequence the dominant social and economic forces in society determine to a large extent what science does and how it does it. More than that, those forces have the power to appropriate from science ideas that are particularly suited to the maintenance and continued prosperity of the social structures of which they are a part. So other social institutions have an input into science both in what is done and how it is thought about, and they take from science concepts and ideas that then support their institutions and make them seem legitimate and natural. It is this dual process—on the one hand, of the social influence and control of what scientists do and say, and, on the other hand, the use of what scientists do and say to further support the institutions of society—that is meant when we speak of science as ideology.

Science serves two functions. First, it provides us with new ways of manipulating the material world by producing a set of techniques, practices, and inventions by which new things are produced and by which the quality of our lives is changed. These are the aspects of science to which scientists appeal when they try to get money from governments or when they appear on the front pages of newspapers in their public relations efforts to maintain their prosperity. We read repeatedly about how "science has discovered" something, but more often than not those announcements are hedged with qualifiers. Biologists discover "evidence for" genes that "may one day" lead to "a possible" cure for cancer. While their over-optimistic reports breed a certain cynicism, it is nevertheless true that scientists do actually change the way we confront the material world.

The second function of science, which is sometimes independent and sometimes closely related to the first, is the function of explanation. Even if scientists are not actually changing the material mode of our existence, they are constantly explaining why things are the way they are. It is often said that these theories about the world must be produced in order, ultimately, to change the world through practice. After all, how can we cure cancer unless we understand what causes cancer? How can we increase food production unless we understand the laws of genetics and plant and animal nutrition?

Yet it is remarkable how much important practical science has been quite independent of theory. In Chapter 3, I will consider one of the most famous examples of scientific agricultural change: the introduction of hybrid corn all over the world. Hybrid corn is said to be one of the great triumphs of modern genetics in action, helping to feed people and increase their well-being. Yet the development of hybrid corn and, indeed, almost all plant and animal breeding as it is actually practiced has been carried out in a way that is completely independent of any scientific theory. Indeed, a great deal of plant and animal breeding has been done in a way indistinguishable from the methods of past centuries before anyone had ever heard of genetics.

The same is true for our attempts to cope with killers like cancer and heart disease. Most cures for cancer involve either removing the growing tumor or destroying it with powerful radiation or chemicals. Virtually none of this progress in cancer therapy has occurred because of a deep understanding of the elementary processes of cell growth and development, although nearly all cancer research, above the purely clinical level, is devoted precisely to understanding the most intimate details of cell biology. Medicine remains, despite all the talk of scientific medicine, essentially an empirical process in which one does what works.

Also in Chapter 3, I will consider the relationship between scientific biology and changes in life expectancy. It is not at all clear that a correct understanding of how the world works is basic to a successful manipulation of the world. But explanations of how the world really works serve another purpose, one in which there has been a remarkable success, irrespective of the practical truth of scientific claims. The purpose is that of *legitimation.*

Regardless of one's political view, everyone must agree that we live in a world in which psychic and material welfare is very unevenly distributed. There are rich people and poor people, sick people and healthy people, people who have control over the conditions of their own lives, work, and time (like professors who are invited to give lectures on the radio and turn them into books) and those who have their tasks assigned to them, who are overseen, who have little or no control over any psychic or material aspect of their lives. There are rich countries and poor countries. Some races dominate others. Men and women have very unequal social and material power.

Some kind of inequality of status, wealth, health, and power have been characteristic of every known society. That means that in every known society there has been some form of struggle between those who have and those who have not, between those with social power and those deprived of it. The uprising of Blacks in America in the 1960s and 1970s, in which there was vast destruction of property and a radical redistribution of consumer goods, and the armed struggle of Mohawks in Canada to prevent the encroachment of commercial and state power on their lands, are only the most recent events in a long history of violent confrontations between those with status, wealth, and power and those without. Repeated peasant uprisings in Europe in the sixteenth and seventeenth centuries resulted in the wholesale destruction of crops and buildings and the loss of hundreds of thousands of lives. The deeds of peasant rebels like Pugachev and Stenka Razin live in song and story. In the United States just after independence from Britain, the farmers of western Massachusetts, led by Daniel Shays and still in possession of their muskets, occupied the general courts to prevent bankers from obtaining judgments to confiscate farmers' property for debt. The bankers in Boston succeeded in getting Continental troops to put down this rebellion, but all at the cost of considerable social upheaval. It is obviously in the interest of those who have power in society to prevent such violent and destructive conflicts, even if, with the police power of the state, they are sure to win.

As such struggles occur, institutions are created whose function is to forestall violent struggle by convincing people that the society in which they live is just and fair, or if not just and fair then inevitable, and that it is quite useless to resort to violence. These are the institutions of social legitimation. They are just as much a part of social struggle as the rick-burnings and machinery destruction of the Captain Swing riots in Britain in the nineteenth century. But they use very different weapons

— ideological weapons. The battleground is in people's heads, and if the battle is won on that ground then the peace and tranquillity of society are guaranteed.

For almost the entire history of European society since the empire of Charlemagne, the chief institution of social legitimation was the Christian Church. It was by the grace of God that each person had an appointed place in society. Kings ruled *Dei gratia*. Occasionally divine grace could be conferred on a commoner who was ennobled, and grace could be removed. Grace was removed from King Charles I, as Cromwell noted, and the proof was Charles's severed head. Even the most revolutionary of religious leaders pressed the claims of legitimacy for the sake of order. Martin Luther enjoined his flock to obey their lords, and in his famous sermon on marriage he asserted that justice was made for the sake of peace and not peace for the sake of justice. Peace is the ultimate social good, and justice is important only if it subserves peace.

For an institution to explain the world so as to make the world legitimate, it must possess several features. First, the institution as a whole must appear to derive from sources outside of ordinary human social struggle. It must not seem to be the creation of political, economic, or social forces, but to descend into society from a supra-human source. Second, the ideas, pronouncements, rules, and results of the institution's activity must have a validity and a transcendent truth that goes beyond any possibility of human compromise or human error. Its explanations and pronouncements must seem to be true in an absolute sense and to derive somehow from an absolute source. They must be true for all time and all place. And finally, the institution must have a certain mystical and veiled quality so that its innermost operation is not completely transparent to everyone. It must have an esoteric language, which needs to be explained to the ordinary person by those who are especially knowledgeable and who can intervene between everyday life and mysterious sources of understanding and knowledge.

The Christian Church or indeed any revealed religion fits

these requirements perfectly, and so religion has been an ideal institution for legitimating society. If only people with special grace, whether they be priests, pastors, or ordinary citizens, are in direct contact with the divine inspiration through revelations, then we must depend upon them completely for an understanding of what has been divinely decreed.

But this description also fits science and has made it possible for science to replace religion as the chief legitimating force in modern society. Science claims a method that is objective and nonpolitical, true for all time. Scientists truly believe that except for the unwanted intrusions of ignorant politicians, science is above the social fray. Theodosius Dobzhansky, a famous scientist who was a refugee from the Bolshevik Revolution and who detested the Bolsheviks, devoted a great deal of energy to pointing out the serious scientific errors that were being made in the Soviet Union in biology and genetics as a consequence of the unorthodox biological doctrines of T.D. Lysenko. It was pointed out to him that, given his own political convictions, he should not carry on that campaign against Lysenko. After all, he believed that sooner or later a global conflict would occur with the United States and the Soviet Union on opposite sides, and he also believed that Lysenko's false scientific doctrines were severely weakening Soviet agricultural production. Why did he then not simply remain quiet about Lysenko's errors so that the Soviet Union would be weakened and compromised in the conflict that was to come? His answer was that his obligation to speak the truth about science was superior to all other obligations and that a scientist must never allow a political consideration to prevent him from saying what he believes to be true.

Not only the methods and institutions of science are said to be above ordinary human relations but, of course, the product of science is claimed to be a kind of universal truth. The secrets of nature are unlocked. Once the truth about nature is revealed, one must accept the facts of life. When science speaks, let no dog bark. Finally, science speaks in mysterious words. No one except an expert can understand what scientists say and do, and we require the mediation of special people — science journalists, for example, or professors who speak on the radio — to explain the mysteries of nature because otherwise there is nothing but indecipherable formulas. Nor can one scientist always understand the formulas of another. Once, when Sir Solly Zuckerman, the famous English zoologist, was asked what he did when he read a scientific paper and came across mathematical formulas, he said, "I hum them."

Despite its claims to be above society, science, like the Church before it, is a supremely social institution, reflecting and reinforcing the dominant values and views of society at each historical epoch. Sometimes the source in social experience of a scientific theory and the way in which that scientific theory is a direct translation of social experience are completely evident, even at a detailed level. The most famous case is Darwin's theory of evolution by natural selection. No scientist doubts that the organisms on earth today have evolved over billions of years from organisms that were very unlike them and that nearly all types of organisms have long since gone extinct. Moreover, we know this to be a natural process resulting from the differential survivorship of different forms. In this sense, we all accept Darwinism as true.

But Darwin's explanation for that evolution is another matter. He claimed that there was a universal struggle for existence because more organisms were born than could survive and reproduce, and that in the course of that struggle for existence, those organisms who were more efficient, better designed, cleverer, and generally better built for the struggle would leave more offspring than the inferior kinds. As a consequence of this victory in the struggle for existence, evolutionary change occurred.

Yet Darwin himself was conscious of the source of his ideas about the struggle for existence. He claimed that the idea for evolution by natural selection occurred to him after reading the famous "Essay on Population" by Thomas Malthus, a late-eighteenth-century parson and economist. The essay was an argument against the old English Poor Law, which Malthus thought too liberal, and in favor of a much stricter control of the poor so they would not breed and create social unrest. In fact, Darwin's whole theory of evolution by natural selection bears an uncanny resemblance to the political economic theory of early capitalism as developed by the Scottish economists. Darwin had some knowledge of the economic survival of the fittest because he earned his living from investment in shares he followed daily in the newspapers. What Darwin did was take early-nineteenth-century *political* economy and expand it to include all of *natural* economy. Moreover, he developed a theory of sexual selection in evolution (about which more will be said in Chapter 4), in which the chief force is the competition among males to be more appealing to discriminating females. This theory was meant to explain why male animals often display bright colors or complex mating dances. It is not clear that Darwin was conscious of how similar his view of sexual selection was to the standard Victorian view of the relationship between middle-class males and females. In reading Darwin's theory, one can see the proper young lady seated on her sofa while the swain on his knees before her begs for her hand, having already told her father how many hundreds a year he has in income.

Most of the ideological influence from society that permeates science is a great deal more subtle. It comes in the form of basic assumptions of which scientists themselves are usually not aware yet which have profound effect on the forms of explanations and which, in turn, serve to reinforce the social attitudes that gave rise to those assumptions in the first place. One of the assumptions is the relation of individual to collectivity, the famous problem of part and whole. Before the eighteenth century, European society placed little or no emphasis on the importance of the individual. Rather, the activity of people was determined for the most part by the social class into which they were born, and individuals confronted each other as representatives of their social group. In a dispute, for example, between a priest and a merchant over a commercial matter, the priest would make his case in an ecclesiastical court and the merchant in the court of his own lord rather than both being subject to the same judgment. Individuals were seen not as the causes of social arrangements but as their consequence.

Moreover, people were not free to move in the economic hierarchy. Peasants and lords alike had mutual obligations and were bound to each other by those obligations. There was no freely moving competitive labor force where each person had the power to sell his or her labor power in a labor market. These relations made it quite impossible to develop the kind of productive capitalism that marks our own era, in which freedom for individuals to move from place to place, from task to task, from status to status, to confront each other sometimes as tenants, sometimes as producers and sometimes as consumers, is an absolute necessity. For example, serfdom had to be abolished in Russia in the middle of the nineteenth century because there was a shortage of factory labor and serfs were legally prohibited from being sent to factories. Sometimes, in fact, serf owners illegally shipped their peasants into factories, and serfs petitioned the czar for relief.

The developing science of the Middle Ages and Renaissance was characterized by seeing all of nature as a kind of indissoluble whole. Living and dead could be transformed one into the other, provided one knew the mystical formula. Nature could not be understood by taking it into pieces because by doing so one destroyed what was essential to it. Alexander Pope said it was "like following life through creatures you dissect./You lose it in the moment you detect." Just as social organization was seen as an indissoluble whole, so was nature.

With the change in social organization that was wrought by developing industrial capitalism, a whole new view of society has arisen, one in which the individual is primary and independent, a kind of autonomous social atom that can move from place to place and role to role. Society is now thought to be the consequence, not the cause, of individual properties. It is individuals who make society. Modern economics is grounded in the theory of consumer preference. Individual autonomous firms compete with each other and replace each other. Individuals have power over their own bodies and labor power, in what MacPherson called "possessive individualism."[1] This atomized society is matched by a new view of nature, the reductionist view. Now it is believed that the whole is to be understood *only* by taking it into pieces, that the individual bits and pieces, the atoms, molecules, cells, and genes, are the causes of the properties of the whole objects and must be separately studied if we are to understand complex nature. Darwin's theory of evolution was a theory of the differential reproductive rate of individuals, and all of the phenomena of evolution were to be understood at this individual causal level. All of modern biology and, indeed, all of modern science takes as its informing metaphor the clock mechanism described by René Descartes in Part V of his *Discourses*. Descartes, being religious, excluded the human soul from the *bête machine*, but that very soon became included as well to make the *homme machine* of the present view. Modern science sees the world, both living and dead, as a large and complicated system of gears and levers.

A second feature of the transformation of scientific views has been the clear distinction between causes and effects. Things are supposed to be either one or the other. Again, in Darwin's view, organisms were acted upon by the environment; they were the passive objects and the external world was the active subject. This alienation of the organism from its outside world means that the outside world has its own laws that are independent of the organisms and so cannot be changed by those organisms. Organisms find the world as it is, and they must either adapt or die. "Nature — love it or leave it." It is the natural analog of the old saw that you can't fight city hall. As I shall show in Chapter 5, this is an impoverished and incorrect view of the actual relationship between organisms and the world they occupy, a world that living organisms by and large *create* by their own living activities.

So, the ideology of modern science, including modern

biology, makes the atom or individual the causal source of all the properties of larger collections. It prescribes a way of studying the world, which is to cut it up into the individual bits that cause it and to study the properties of these isolated bits. It breaks the world down into independent autonomous domains, the internal and the external. Causes are either internal or external, and there is no mutual dependency between them.

For biology, this world view has resulted in a particular picture of organisms and their total life activity. Living beings are seen as being determined by internal factors, the genes. Our genes and the DNA molecules that make them up are the modern form of grace, and in this view we will understand what we are when we know what our genes are made of. The world outside us poses certain problems, which we do not create but only experience as objects. The problems are to find a mate, to find food, to win out in competition over others, to acquire a large part of the world resources as our own, and if we have the right kinds of genes we will be able to solve the problems and leave more offspring. So in this view, it is really our genes that are propagating themselves through us. We are only their instruments, their temporary vehicles through which the self-replicating molecules that make us up either succeed or fail to spread through the world. In the words of Richard Dawkins, one of the leading proponents of this biological view, we are "lumbering robots" whose genes "created us body and mind."

Just as at one level genes determine individuals, so at another level it is individuals who determine collectivities. If we want to understand why an ant colony has a particular division of tasks or a flock of birds flies in a particular way, we need only look at the individual ants and individual birds, because the behavior of the group is a consequence of the behaviors of the individual organisms; that behavior is in turn determined by genes. For human beings that means that the structure of our society is nothing but a result of the collection of individual behaviors. If our country goes to war, we are told it is because we feel aggressive as individuals. If we live in a competitive entrepreneurial society, it is because, in this view, each one of us, as an individual, has a drive to be competitive and entrepreneurial.

Genes make individuals and individuals make society, and so genes make society. If one society is different from another, that is because the genes of the individuals in one society are different from those in another. Different races are thought to be genetically different in how aggressive or creative or musical they are. Indeed, culture as a whole is seen as made up of little bits and pieces of cultural bric-a-brac, what some sociobiologists call *culturgens*. In this view, a culture is a sack of bits and pieces such as aesthetic preferences, mating preferences, work and leisure preferences. Dump out the sack and culture will be displayed before you. Thus, the hierarchy is complete. Genes make individuals, individuals have particular preferences and behaviors, the collection of preferences and behaviors makes a culture, and so genes make culture. That is why molecular biologists urge us to spend as much money as necessary to discover the sequence of the DNA of a human being. They say that when we know the sequence of the molecule that makes up all our genes, we will know what it is to be human. When we know what our DNA looks like, we will know why some of us are rich and some poor, some healthy and some sick, some powerful and some weak. We will also know why some societies are powerful and rich and others are weak and poor, why one nation, one sex, one race dominates another. Indeed, we will know why there is such a thing as a science of biology, which itself is one of the bits and pieces of culture lying at the bottom of the sack.

We have become so used to the atomistic machine view of the world that originated with Descartes that we have forgotten that it is a metaphor. We no longer think, as Descartes did, that the world is *like* a clock. We think it *is* a clock. We cannot imagine an alternative view unless it be one that goes back to a prescientific era. For those who are dissatisfied with the modern world and dislike the artifacts of science, the pollution, the noise, the industrial world, the overmechanized medical care that seems not to make us feel better much of the time — for people who want to go back to nature and the good old ways, the response has been to return to a description of the world as an indissoluble whole that we murder to dissect. For them, there is no use in trying to break anything down into parts because we inevitably lose the essence, and the best we can do is treat the world holistically.

But this holistic world view is untenable. It is simply another form of mysticism and does not make it possible to manipulate the world for our own benefit. An obscurantist holism has been tried and it has failed. The world is not one huge organism that regulates itself to some good end as the believers in the Gaia hypothesis believe. While in some theoretical sense "the trembling of a flower is felt on the farthest star," in practice my gardening has no effect on the orbit of Neptune because the force of gravitation is extremely weak and falls off very rapidly with distance. So there is clearly truth in the belief that the world can be broken up into independent parts. But that is not a universal direction for the study of all nature. A lot of nature, as we shall see, cannot be broken up into independent parts to be studied in isolation, and it is pure ideology to suppose that it can.

The problem is to construct a third view, one that sees the entire world neither as an indissoluble whole nor with the equally incorrect, but currently dominant, view that at every level the world is made up of bits and pieces that can be isolated and that have properties that can be studied in isolation. Both ideologies, one that mirrors the premodern feudal social world, and the other that mirrors the modern competitive individualist entrepreneurial one, prevent us from seeing the full richness of interaction in nature. In the end, they prevent a rich understanding of nature and prevent us from solving the problems to which science is supposed to apply itself.

In the ensuing chapters, we will look in some detail at particular manifestations of the modern scientific ideology and the false paths down which it has led us. We will consider how biological determinism has been used to explain and justify inequalities within and between societies and to claim that those inequalities can never be changed. We will see how a theory of human nature has been developed using Darwin's theory of evolution by natural selection to claim that social organization is also unchangeable because it is natural. We will see how problems of health and disease have been located within the individual so that the individual becomes a problem for society to cope with rather than society becoming a problem for the individual. And we will see how simple economic relationships masquerading as facts of nature can drive the entire direction of biological research and technology.

While these examples are meant to disillusion the reader about the objectivity and vision of transcendent truth claimed by scientists, they are not intended to be antiscientific or to suggest that we should give up science in favor of, say, astrology or thinking beautiful thoughts. Rather, they are meant to acquaint the reader with the truth about science as a social activity and to promote a reasonable skepticism about the sweeping claims that modern science makes to an understanding of human existence. There is a difference between skepticism and cynicism, for the former can lead to action and the latter only to passivity. So these pages have a political end, too, which is to encourage the readers not to leave science to the experts, not to be mystified by it, but to demand a sophisticated scientific understanding in which everyone can share.

ALL IN THE GENES?

*O*ur society was born, at least politically, in revolutions of the seventeenth century in Britain and the eighteenth century in France and America. Those revolutions swept out an old order characterized by aristocratic privilege and a relative fixity of persons in the society. The bourgeois revolutions in England, France, and America claimed that this old society and its ideology were illegitimate, and the ideologues of those revolutions produced and legitimized an ideology of liberty and equality. Diderot and the Encyclopedists and Tom Paine were the theorists of a society of "liberté, égalité, fraternité," of all men created equal. The writers of the Declaration of Independence asserted that political truths were "self-evident; that all men are created equal; that they are endowed by their creator with certain unalienable rights; that among these are life, liberty, and the pursuit of happiness" (by which, of course, they meant the pursuit of money). They meant literally all *men*, because women were not given the right to vote in the United States until 1920; Canada enfranchised women a little sooner, in 1918—but not in provincial elections in Quebec until 1940. And of course they didn't mean *all* men, because slavery continued in the French dominions and in the Caribbean until the middle of the nineteenth century. Blacks were defined by the United States Constitution as only three-fifths of a person, and for most of the history of English parliamentary democracy, a man had to have money to vote.

To make a revolution, you need slogans that appeal to the great mass of people, and you could hardly get people to shed blood under a banner that read "Equality for some." So the ideology and the slogans outstrip the reality. For if we look at the society that has been created by those revolutions, we see a great deal of inequality of wealth and power among individuals, between sexes, between races, between nations. Yet we have heard over and over again in school and had it drummed into us by every organ of communication that we live in a society of free equals. The contradiction between the claimed equality of

our society and the observation that great inequalities exist has
been, for North Americans at least, the major social agony of
the last 200 years. It has motivated an extraordinary amount of
our political history. How are we to resolve the contradiction of
immense inequalities in a society that claims to be founded on
equality?

There are two possibilities. We might say that it was all a fake,
a set of slogans meant to replace a regime of aristocrats with a
regime of wealth and privilege of a different sort, that inequality
in our society is structural and an integral aspect of the whole of
our political and social life. To say that, however, would be
deeply subversive, because it would call for yet another
revolution if we wanted to make good on our hopes for liberty
and equality for all. It is not a popular idea among teachers, newspaper editors, college professors, successful politicians, indeed
anyone who has the power to help form public consciousness.

The alternative, which has been the one taken since the
beginning of the nineteenth century, has been to put a new gloss
on the notion of equality. Rather than equality of *result*, what
has been meant is equality of *opportunity*. In this view of equality, life is a foot race. In the bad old days of the *ancien régime*,
the aristocrats got to start at the finish line whereas all the rest of
us had to start at the beginning, so the aristocrats won. In the new
society, the race is fair: everyone is to begin at the starting line
and everyone has an equal opportunity to finish first. Of course,
some people are faster runners than others, and so some get the
rewards and others don't. This is the view that the old society
was characterized by *artificial* barriers to equality, whereas the
new society allows a natural sorting process to decide who is to
get the status, wealth, and power and who is not.

Such a view does not threaten the status quo, but on the contrary supports it by telling those who are without power that their
position is the inevitable outcome of their own innate deficiencies
and that, therefore, nothing can be done about it. A remarkably
explicit recent statement of this assertion is the one by Richard
Herrnstein, a psychologist from Harvard, who is one of the most

outspoken modern ideologues of natural inequality. He wrote,

> the privileged classes of the past were probably not much superior biologically to the downtrodden which is why revolution had a fair chance of success. By removing artificial barriers between classes society has encouraged the creation of biological barriers. When people can take their natural level in society, the upper classes will, by definition, have greater capacity than the lower.[1]

We are not told precisely what principle of biology guarantees that biologically inferior persons cannot seize power from biologically superior ones, but it is not logic that is at issue here. Such statements as Herrnstein's are meant to convince us that although we may not live in the best of all *conceivable* worlds, we live in the best of all *possible* worlds. The social entropy has been maximized so that we have as much equality as possible because the structure is essentially one of equality, and whatever inequalities are left over are not structural but based on innate differences between individuals. In the nineteenth century this was also the view, and education was seen as the lubricant that would guarantee that the race of life was run smoothly. Lester Frank Ward, a giant of nineteenth-century sociology, wrote, "Universal education is the power which is destined to overthrow every species of hierarchy. It is destined to remove all artificial inequality and leave the natural inequalities to find their true level. The true value of a newborn infant lies in its naked capacity for acquiring the ability to do."[2]

This was echoed 60 years later by Arthur Jensen at the University of California, who wrote about the inequality of intelligence of Blacks and whites: "We have to face it, the assortment of persons into occupational roles simply is not fair in any absolute sense. The best we can hope for is that true merit given equality of opportunity acts as a basis for the natural assorting process."[3]

Simply to assert that the race of life is fair and that different people have different intrinsic abilities to run it is not enough to explain the observations of inequality. Children seem, by and large, to acquire the social status of their parents. About 60 percent of the children of "blue collar" workers remain "blue collar," while about 70 percent of "white collar" workers' children are "white collar." But these figures vastly overestimate the amount of social mobility. Most people who have passed from "blue collar" to "white collar" jobs have passed from factory production-line jobs to office production-line jobs or have become sales clerks, less well paid, less secure, doing work just as numbing of the soul and body as the factory work done by their parents. The children of gas station attendants usually borrow money, and the children of oil magnates usually lend it. The chance that Nelson Rockefeller would have wound up pumping gas was pretty close to zero.

If we live in a meritocracy, in which each person can rise to the status allowed by his or her innate capacities, how do we explain this passage of social power from parent to offspring? Are we really just back in an old aristocratic situation? The naturalistic explanation is to say that not only do we differ in our innate capacities but that these innate capacities are themselves transmitted from generation to generation biologically. That is to say, they are in our genes. The original social and economic notion of inheritance has been turned into biological inheritance.

But even the claim that the intrinsic ability to win success is inherited in the genes is not sufficient to justify an unequal society. After all, we might assert that there ought not to be any particular relationship between what one can accomplish and what social and psychic rewards are given. We might give the same material and psychic rewards to house painters and picture painters, to surgeons and to barbers, to professors who give lectures, and to the janitors who come in and clean up the classroom afterward. We might create a society on whose banners are inscribed, "From each according to his ability, to each according to his need."

To meet this objection to an unequal society there has been developed a biological theory of human nature that says that while the differences between us are in our genes, there are certain inborn similarities among us all. These similarities of human nature guarantee that differences in ability will be converted into differences in status, that society is naturally hierarchical, and that a society of equal reward and status is biologically impossible. We might pass laws requiring such equality, but the moment the vigilance of the state was relaxed we would return to "doing what comes naturally."

These three ideas—that we differ in fundamental abilities because of innate differences, that those innate differences are biologically inherited, and that human nature guarantees the formation of a hierarchical society—when taken together, form what we can call the *ideology of biological determinism*.

The idea that blood will tell was not invented by biologists. It is a dominant theme of nineteenth-century literature, and one can hardly appreciate the most praised and popular writers of the last century without seeing how a theory of innate difference informed their work. Think of Dickens's *Oliver Twist*. When Oliver first meets young Jack Dawkins, the Artful Dodger, on the road to London, a remarkable contrast in body and spirit is established. The Dodger is described as "a snub-nosed, flat-browed, common-faced boy ... with rather bow-legs, and little, sharp, ugly eyes," and his English was not the best. What can we expect from a 10-year-old street urchin with no family, no education, and only the lowest criminals of London for companions? Oliver's speech, however, is perfect (he knows when to use the subjunctive) and his manner is genteel. He is described as a pale, thin child, but with a good sturdy spirit in his breast. Yet Oliver was raised from birth in the most degrading of nineteenth-century British institutions, the parish workhouse, an orphan with no education and little to eat. He is described as having spent the first nine years of his life rolling about on the floor all day "without the inconvenience of too much food or too much clothing." Where amid the oakum-pickings did Oliver garner that sensitivity of soul and perfection of English grammar? *Oliver Twist* is a mystery novel, and that is its mystery. The answer is that although his food was gruel, his blood was upper-middle-class. His mother was the daughter of a naval officer. His father's family was well off and socially ambitious.

A similar theme is central to George Eliot's Daniel Deronda. We first meet Daniel, the young stepson of an English baronet, wasting his time in a fashionable gambling spa. When he becomes a bit older, he suddenly has mysterious longings for things Hebrew. He falls in love with a Jewish woman, studies the Talmud, and converts. The reader will not be surprised to learn that he is the son of a Jewish actress whom he has never seen but whose blood tells. Nor is this a madness only of the Anglo-Saxons. The Rougon-Macquart novels of Émile Zola were deliberately written as a kind of experimental literature to illustrate the discoveries of nineteenth-century anthropology. In the preface, Zola tells us that "heredity has its laws just like gravitation." The Rougon-Macquarts are a family descended from the two lovers of one woman, one of whom was a solid, industrious peasant, while the other was a wastrel and degenerate. From the dependable peasant descend solid, honest stock, while from the degenerate ancestor descend a long line of social misfits and criminals including the famous Nana, who was a nymphomaniac from early childhood, and her mother, Gervaise, the laundress, who despite beginning a solid entrepreneurial life, lapses into her natural indolence. When Gervaise's husband, Copeau, the father of Nana, was admitted to hospital with the D.T.s, the first question the physician asked him was, "Did your father drink?" The public consciousness of the period both in Europe and North America was permeated with the notion that intrinsic differences in temperament and merit will finally dominate any mere effect of education and environment.

The fictional Rougon-Macquarts are seen again in the equally fictional but supposedly real family of Kallikaks, who graced virtually every textbook of American psychology until the Second World War. The Kallikaks were supposed to be two halves of a family descended from two women of contrasting nature and a common father. This piece of academic fiction was meant to convince malleable young minds that criminality, laziness, alcoholism, and incest were inborn and inherited.

Nor were supposedly innate differences restricted to individual variation. Nations and races were said to be characterized by innate temperamental and intellectual differences. These claims were made not by racists, demagogues, and fascist know-nothings but by the leaders of the American academic, psychological, and sociological establishments. In 1923, Carl Brigham, who was later secretary of the College Entrance Examination Board, produced a study of intelligence under the direction of R.M. Yerkes, professor of psychology at Harvard and the president of the American Psychological Association. The study asserted: "We must assume that we are measuring inborn intelligence. We must face the possibility of racial admixture here in America that is infinitely worse than that faced by any European country for we are incorporating the Negro into our racial stock. The decline of the American intelligence will be more rapid... owing to the presence here of the Negro."[4]

Yet another president of the American Psychological Association said that whenever there has been mixed breeding with the Negro, there has been deterioration of civilizations.[5] Louis Agassiz, one of the most famous zoologists of the nineteenth century, reported that the skull sutures of Negro babies closed earlier than the sutures of white babies, so their brains were entrapped, and it would be dangerous to teach them too much. Perhaps the most extraordinary of claims was that of Henry Fairfield Osborne, president of the American Museum of Natural History and one of America's most eminent and prestigious pale-ontologists, who worked out the sequence of evolution of the horse. He wrote,

> The northern races invaded the countries to the south, not only as conquerors but as contributors of strong moral and intellectual elements to a more or less decadent civilization. Through the Nordic tide which flowed into Italy came the ancestors of Raphael, Leonardo, Galileo, Titiano; also, according to

Günther, of Giotto, Botticelli, Petrarca, and Tasso. Columbus, from his portraits and from busts, *whether authentic or not*, was clearly of Nordic ancestry.[6] [emphasis added]

Whether authentic or not, indeed! Over and over again, leading intellectuals have assured their audiences that modern science shows that there are inborn racial and individual differences in ability. Nor have modern biologists taken a different view. Except for a brief interruption around the time of the Second World War, when the crimes of Nazism made claims of innate inferiority extremely unpopular, biological determinism has been the mainstream commitment of biologists. Yet these claims are made without a shred of evidence and in contradiction to every principle of biology and genetics.

To realize the error of these claims, we need to understand what is involved in the development of an organism. First, we are not determined by our genes, although surely we are influenced by them. Development depends not only on the materials that have been inherited from parents—that is, the genes and other materials in the sperm and egg—but also on the particular temperature, humidity, nutrition, smells, sights, and sounds (including what we call education) that impinge on the developing organism. Even if I knew the complete molecular specification of every gene in an organism, I could not predict what that organism would be. Of course, the difference between lions and lambs is almost entirely a consequence of the difference in genes between them. But variations among individuals within species are a unique consequence of both genes and the developmental environment in a constant interaction. Moreover, curiously enough, even if I knew the genes of a developing organism and the complete sequence of its environments, I could not specify the organism.

There is yet another factor at work. If we count the number of bristles under the wing of a fruitfly, for example, we find that there is a different number on the left side than on the right. Some have more bristles on the left, some more on the right; there is no average difference. So, there is a kind of fluctuating asymmetry. An individual fruitfly, however, has the same genes on its left side as on its right. Moreover, the tiny size of a developing fruitfly and the place it develops guarantee that both left and right sides have had the same humidity, the same oxygen, the same temperature. The differences between left and right side are caused neither by genetic nor by environmental differences but by random variation in growth and division of cells during development: *developmental noise.*

This chance element in development is an important source of variation. Indeed, in the case of the fruitfly bristles, there is as much variation consequent on developmental noise as there is from genetic and environmental variation. We do not know in human beings, for example, how much of the difference between us is a consequence of the random differences in the growth of neurons during our embryonic life and early childhood. It is our common prejudice that even if one had practiced the violin from a very early age, one would not be able to play as well as Menuhin, and we think of him as having special neuronal connections. But that is not the same as saying that those neuronal connections were coded in his genes. There may be large random differences in the growth of our central nervous systems. It is a fundamental principle of developmental genetics that every organism is the outcome of a unique interaction between genes and environmental sequences modulated by the random chances of cell growth and division, and that all these together finally produce an organism. Moreover, an organism changes throughout its entire life. Human beings change their size, not only growing larger as children, but as they grow old, growing smaller as their joints and bones shrink.

A more sophisticated version of genetic determinism agrees that organisms are a consequence of both environmental and genetic influences but describes differences between individuals as differences in *capacity.* This is the empty bucket metaphor. We each begin life as an empty bucket of a different size. If the environment provides only a little water, then all these buckets will have the same amount in them. But if an abundance is provided from the environment, then the small buckets will overflow and the large ones will hold more. In this view, if every person were allowed to develop to his or her genetic capacity, there would indeed be major differences in ability and performance, and these would be fair and natural.

But there is no more biology in the metaphor of innate capacity than there is in the notion of fixed genetic effects. The unique interaction between organism and environment cannot be described by differences in capacity. It is true that if two genetically different organisms developed in exactly the same environment, they would be different, but that difference cannot be described as different capacities because the genetical type that was superior in one environment may be inferior in a second developmental environment. For example, strains of rats can be selected for better or poorer ability to find their way through a maze, and these strains of rats pass on their differential ability to run the maze to their offspring, so they are certainly genetically different in this respect. But if exactly the same strains of rats are given a different task, or if the conditions of learning are changed, the bright rats turn out to be dull and the dull rats turn out to be bright. There is no general genetic superiority of one rat strain over another in finding its way through a problem.

A more subtle and mystifying approach to biological determinism rejects both the genetic fixity of the first view and the capacity metaphor of the second and is, instead, statistical. Essentially, it states the problem as one of partitioning the effects of environment and genes so that we can say that, perhaps, 80 percent of the difference among individuals is caused by their genes and 20 percent by their environment. Of course, these differences must be on a population level rather than an individual level. It would make no sense at all to say that of someone's height of five feet eleven and a half inches, five feet two were a result of her genes and the other nine and a half inches were put there by the food she ate. The statistical view considers the proportion of *variation* among individuals rather than partitioning a particular individual measurement. The statistical approach tries to assign some proportion of all of the variation among individuals or groups to variation among their genes, and a second proportion that results from variation among their environments.

The implication is that if most of the variation in, say, intelligence among individuals is a consequence of variation among their genes, then manipulating the environment will not make much difference. It is often said, for example, that 80 percent of the variation among individual children in their I.Q. performance is caused by variation in their genes and only 20 percent by variation in their environments. The result is that the greatest possible amelioration of environment could not eliminate more than 20 percent of differences among individuals, and the 80 percent would still be there because it is a consequence of genetic variation. This is a completely fallacious although plausible-sounding argument. There is no connection whatsoever between the variation that can be ascribed to genetic differences as opposed to environmental differences and whether a change in environment will affect performance and by how much. We should remember that any very ordinary arithmetic student in primary school in Canada can correctly add a column of figures vastly more quickly than the most intelligent Ancient Roman mathematician, who had to struggle with cumbersome X's, V's, and I's. That same ordinary student can multiply two five-digit numbers with a $10 hand-held calculator more quickly and accurately than a professor of mathematics could have a century ago.

A change in environment, in this case of cultural environment, can change abilities by many orders of magnitude. Moreover, the differences between individuals are abolished by cultural and mechanical inventions. Differences that can be ascribed to genetic differences and that appear in one environment may disappear completely in another. Although there may be biologically based average differences in physique and strength between a random group of men and random group of women (and these are less than usually supposed), these differences rapidly become irrelevant and disappear from practical view in a world of electrically driven hoists, power steering, and electronic controls. So the proportion of variation in a population as a consequence of variation in genes is not a fixed property but one that varies from environment to environment. That is, how much difference among us is a consequence of genetic differences between us depends, curiously enough, on environment.

Conversely, how much difference there is between us that is a consequence of environmental variation in our life histories depends on our genes. We know from experiments that organisms that have some particular genes are very sensitive to environmental variation while other individuals with different genes are insensitive to environmental variation. Environmental variation and genetic variation are not independent causal pathways. Genes affect how sensitive one is to environment, and environment affects how relevant one's genetic differences may be. The interaction between them is indissoluble, and we can separate genetic and environmental effects statistically only in a particular population of organisms at a particular moment with a particular set of specified environments. When an environment changes, all bets are off.

The contrast between genetic and environmental, between nature and nurture, is not a contrast between fixed and changeable. It is a fallacy of biological determinism to say that if differences are in the genes, no change can occur. We know this to be true from medical evidence alone. There are many so-called inborn errors of metabolism in which a defective gene results, in normal circumstances, in a defective physiology. An example is Wilson's disease, a genetic defect that prevents its sufferers from detoxifying the copper that we all consume in minute quantities in our ordinary food. The copper builds up in the body and eventually causes nervous degeneration and finally death, some time in adolescence or early adulthood. Nothing could be more perfectly described as a genetic disorder. Yet people with this defective gene can lead a perfectly normal life and have a normal development by taking a pill that helps them get rid of the copper, and they are then indistinguishable from anyone else.

It is sometimes said that examples of changing the conditions of performance, such as the invention of Arabic numerals, or the calculator, or providing a pill, are beside the point because we are interested in some sort of basic unaided, naked ability. But there are no measures of "unaided" ability, nor are we really interested in them. There are some people who can remember long columns of figures and others who are good at adding and multiplying large numbers in their heads. So why do we give written I.Q. tests, which, after all, are simply giving the crutch of paper and pencil to people who do not have the "unaided" ability to do mental arithmetic? Indeed, why do we allow people taking mental tests to wear eyeglasses, if we are interested in culturally unmodified "naked" abilities? The answer is that we have no interest in arbitrarily defined abilities, but are concerned with differences in the ability to carry out *socially constructed* tasks that are relevant to the structure of our actual social lives.

Aside from the conceptual difficulties of trying to ascribe separate effects to genes and environment, there are severe experimental difficulties in detecting the influence of genes, especially when we deal with human beings. How do we decide whether genes influence differences in some trait? In all organisms the process is the same. We compare individuals who are differently related to one another, and if more closely related individuals are more similar than are more distantly related ones, we ascribe some power to the genes. But herein lies the deep difficulty of human genetics. Unlike experimental animals, people who are more closely related to each other not only share more genes in common but they also share environment in common because of the family and class structure of human societies. The observation that children resemble their parents in some trait does not distinguish between similarity that comes from genetic similarity and similarity that arises from environmental resemblance. The resemblance of parents and children is the observation to be explained. It is not evidence for genes. For example, the two social traits that have the highest resemblance between parents and children in North America are religious sect and political party. Yet even the most ardent biological determinist would not seriously argue that there is a gene for Episcopalianism or voting Social Credit.

The problem is to distinguish genetic similarity from environmental similarity. It is for this reason that so much emphasis has been put on twin studies in human genetics. The idea is that if twins are more similar than ordinary sibs or if twins raised in completely isolated families are still similar, then this surely must be evidence for genes. In particular, there has been a fascination with a study of identical twins raised apart. If identical twins—that is, twins sharing all the same genes—are similar even though raised apart, then their traits must be strongly genetically influenced. Much of the claim for the high heritability of I.Q., for example, comes from studies of identical twins raised apart.

Only three such studies have been published. The first and largest set of studies was reported by Sir Cyril Burt. This was the only study that claimed no similarity between the family circumstances of the families that raised separated twins. It also claimed a heritability of 80 percent for I.Q. performance. However, careful investigation by Oliver Gillie of the *Times* of London and Professor Leon Kamin at Princeton revealed that Burt had simply made up the numbers and made up the twins.[7] He even made up the collaborators whose names appeared with his in the publications. We need consider these claims no further. They represent one of the great scandals of modern psychology and biology.

When we look at the other studies, which actually give family details of the separated twins, we realize that we live in a real world and not in a Gilbert and Sullivan operetta. The reason that twins are separated at birth may be that their mother has died in childbirth, so that one twin is raised by an aunt and another by a best friend or grandmother. Sometimes the parents cannot afford to keep both children so they give one to a relative. In fact, the studied twins were not raised apart at all. They were raised by members of the same extended family, in the same small village. They went to school together. They played together. Other adoption studies of human I.Q. that are said to demonstrate the effect of genes have their own experimental difficulties, including the failure to match children by age, extremely small samples, and biased selection of cases for study.[8] There is a strong effort on the part of parents of many twins to make them as similar as possible. They are given names beginning with the same letter and are dressed alike. International twin conventions give prizes for the most similar twins. One twin study advertised in the newspapers and offered a free trip to Chicago for identical twins, thus attracting those who were the most similar.[9] As a consequence of such biases, there is at present simply no convincing measure of the role of genes in influencing human behavioral variation.

One of the major biological ideological weapons used to convince people that their position in society is fixed and unchangeable and, indeed, fair is the constant confusion between inherited and unchangeable. This confusion is nowhere more manifest than in the very studies of adoptions that are meant to measure biological similarities. In human populations, one carries out an adoption study like that of separation of identical twins to try to break the connection between resemblance that comes from genetic sources and resemblance that comes from the sources of family similarity. If adopted children resemble their biological parents more closely than they resemble their adopting parents, then the geneticists quite correctly regard this as evidence for the influence of genes. When one looks at all the studies of adoption in order to study the genetic influence on intelligence, there are two constant results.

First, adopted children do resemble their biological parents in the sense that the higher the I.Q. score of the biological parent, the higher the I.Q. score of the child who was adopted. So, biological parents are having some influence on the I.Q. of their children even though those children are adopted early, and putting aside the possibility of prenatal nutritional differences or extremely early stimulation, it would be reasonable to say that genes have some influence on I.Q. scores. We can only speculate about the source of genetic influence. There is a premium on speed in I.Q. testing, and genes might have some influence on reaction times or general speed of central nervous processes.

The second feature of adoption studies is that the I.Q. test scores of the children are about 20 points higher than those of their biological parents. It is still the case that the biological parents with the higher I.Q. scores have children with higher scores, but the children as a group have moved well ahead of their biological parents. In fact, the average I.Q. scores of these adopted children are about equal to the average I.Q. of the adopting parents, who always do much better on I.Q. tests than the biological parents. What is at stake here is the difference between *correlation* and *identity*. Two variables are positively correlated if higher values of one are matched with higher values of the other. The ordered set of numbers 100, 101, 102, and 103 is perfectly correlated with the set of 120, 121, 122, and 123 because each increase in one set is perfectly matched by an increase in the other. Yet the two sets of numbers are clearly not identical, differing as they do by 20 units on the average. So the I.Q. of parents may be excellent predictors of the I.Q.s of their children in the sense that higher values for parents are matched with higher values for offspring, but the *average* I.Q. value of their children may be much greater. For the geneticist, it is the *correlation* that indicates the role of genes; the heritability predicts nothing about changes in the group average from generation to generation. The adoption studies are a revelation of the meaning of I.Q. tests and of the social reality of adoption.

First, what do I.Q. tests actually measure? They are a combination of numerical, vocabulary, educational, and attitudinal questions. They ask such things as "Who was Wilkins McCawber?", "What is the meaning of 'sudiferous'?", "What should a girl do if a boy hits her?" (Hitting him back is *not* the right answer!) And how do we know that someone who does well on such a test is *intelligent*? Because, in fact, the tests were originally standardized to pick out precisely those children in a class whom the teacher had already labeled intelligent. That is, I.Q. tests are instruments for giving an apparently objective and "scientific" gloss to the social prejudices of educational institutions.

Second, people who decide on an early adoption for their children are usually working-class or unemployed people who do not share in the education and culture of the middle class. People who adopt children, on the other hand, are usually middle-class and have an appropriate education and cultural experience for the content and intent of I.Q. tests. So adopting parents have, as a group, much higher I.Q. performances than the parents who have chosen adoption for their children. The educational and family environment in which these children are then raised has the expected result of raising all their I.Q.s even though there is evidence for some genetic influence from their biological parents.

These results of adoption studies illustrate perfectly why we cannot answer a question about how much something can be changed by answering a different question, namely, are there genes influencing the trait? If we wanted seriously to ask the question posed by Arthur Jensen in his famous article "How much can we boost I.Q. and scholastic achievement?",[10] the only way we could answer would be to try to boost I.Q. and scholastic achievement. We do not answer it by asking, as Jensen did, whether there is a genetic influence on I.Q., because to be genetic is not to be unchangeable.

Biological determinists claim that there are not only differences in ability among individuals but that these individual differences explain racial differences in social power and success. It is hard to know how one would get evidence about Black–white differences that did not totally confound genetic and environmental variation. Inter-racial adoptions, for example, are uncommon, especially of white children adopted by Black foster parents. Occasional evidence does appear, however.

In Dr. Bernardo's homes in Britain, where children are taken as orphans soon after birth, a study was done of intelligence testing of children of Black and white ancestry.[11] Several tests were given at various ages, and small differences were found in the I.Q. performance between these groups, but these were not statistically significant. If nothing more was said about it, most of the readers would assume that the small differences showed whites were better than Blacks. But in fact the reverse was true. The differences were not statistically significant, but where there were any differences, they were in favor of Blacks. There is not an iota of evidence of any kind that the differences in status, wealth, and power between races in North America have anything to do with the genes, except, of course, for the socially mediated effects of the genes for skin color. Indeed, there is in general a great deal less difference genetically between races than one might suppose from the superficial cues we all use in distinguishing races. Skin color, hair form, and nose shape are certainly influenced by genes, but we do not know how many such genes there are, or how they work. On the other hand, when we look at genes we do know something about, genes that influence our blood type, for example, or genes for the various enzyme molecules essential to our physiology, we find that although there is a tremendous amount of variation from individual to individual, there is remarkably little variation on the average between major human groups. In fact, about 85 percent of all identified human genetic variation is between any two individuals from the same ethnic group. Another 8 percent of all the variation is between ethnic groups within a race—say, between Spaniards, Irish, Italians, and Britons—and only 7 percent of all human genetic variation lies on the average between major human races like those of Africa, Asia, Europe, and Oceania.[12]

So we have no reason *a priori* to think that there would be any genetic differentiation between racial groups in characteristics such as behavior, temperament, and intelligence. Nor is there an iota of evidence that social classes differ in any way in their genes except insofar as ethnic origin or race may be used as a form of economic discrimination. The nonsense propagated by ideologues of biological determinism that the lower classes are biologically inferior to the upper classes, that all the good things in European culture come from the Nordic groups, is precisely nonsense. It is meant to legitimate the structures of inequality in our society by putting a biological gloss on them and by propagating the continual confusion between what may be influenced by genes and what may be changed by social and environmental alterations.

The vulgar error that confuses heritability and fixity has been, over the years, the most powerful single weapon that biological ideologues have had in legitimating a society of inequality. Since as biologists they must know better, one is entitled to at least a suspicion that the beneficiaries of a system of inequality are not to be regarded as objective experts.

CAUSES AND
THEIR EFFECTS

*M*odern biology is characterized by a number of ideological prejudices that shape the form of its explanations and the way its researches are carried out. One of those major prejudices is concerned with the nature of causes. Generally one looks for *the* cause of an effect, or even if there are a number of causes allowed, one supposes that there is a major cause and the others are only subsidiary. And in any case, these causes are separated from each other, studied independently, and manipulated and interfered with in an independent way. Moreover, these causes are usually seen to be at an individual level, the individual gene or the defective organ or an individual human being who is the focus of internal biological causes and external causes from an autonomous nature.

This view of causes is nowhere more evident than in our theories of health and disease. Any textbook of medicine will tell us that the cause of tuberculosis is the tubercle bacillus, which gives us the disease when it infects us. Modern scientific medicine tells us that the reason we no longer die of infectious diseases is that scientific medicine, with its antibiotics, chemical agents, and high-technology methods of caring for the sick, has defeated the insidious bacterium.

What is the cause of cancer? The cause is the unrestricted growth of cells. That runaway growth, in turn, is a consequence of the failure of certain genes to regulate cell division. So we get cancer because our genes are not doing their business. It used to be that people thought that viruses were a major cause of cancer, and a great deal of money and time has been spent looking for the viral causes of cancer in humans without success. Biology has moved on from the time when viruses were all the rage to a time when genes are much more trendy.

Alternatively, there are environmental insult theories of the causes of cancer. Cancers are caused, we are told, by asbestos or by PVC or by a host of natural chemicals over which we have no control, and although they are present in very low concentrations, we are exposed to them over our whole lives. So, just as we will avoid dying from tuberculosis by dealing with the bug that causes it, so we will avoid dying from cancer by getting rid of particularly nasty chemicals in the environment. It is certainly true that one cannot get tuberculosis without a tubercle bacillus, and the evidence is quite compelling that one cannot get the cancer mesothelioma without having ingested asbestos or related compounds. But that is not the same as saying that *the* cause of tuberculosis is *the* tubercle bacillus and *the* cause of mesothelioma is asbestos. What are the consequences for our health of thinking in this way? Suppose we note that tuberculosis was a disease extremely common in the sweatshops and miserable factories of the nineteenth century, whereas tuberculosis rates were much lower among country people and in the upper classes. Then we might be justified in claiming that *the* cause of tuberculosis is unregulated industrial capitalism, and if we did away with that system of social organization, we would not need to worry about the tubercle bacillus. When we look at the history of health and disease in modern Europe, that explanation makes at least as good sense as blaming the poor bacterium.

What is the evidence for the benefits of modern scientific medicine? Certainly we live a great deal longer than our ancestors. In 1890, the years of life expected for a white child at birth in North America were only 45, whereas now the expected life span is 75 years, but that is not because modern medicine has prolonged the life of elderly and sick people. A very large fraction of the change in the average life expectancy is a tremendous reduction in infant mortality. Before the turn of the century and especially earlier in the nineteenth century, there was a considerable chance that a child never got to be a year old—in 1860, the infant mortality rate in the U.S. was 13 percent—, so the average life expectancy for the population as a whole was reduced considerably by this early death. The gravestones of people who died in the middle of the nineteenth century indicate a remarkable number of deaths at an old age. In fact, scientific medicine has done little to add years for people who have already reached their maturity. In the last 50 years, only about four months have been added to the expected life span of a person who is already 60 years old.

As we all know, in modern Europe women live longer than men, but they used not to. Before the turn of the century, women died sooner than men did, and a common explanation offered by scientific medicine is that a leading cause of death in women in the days before modern medicine was childbirth fever. According to this view, modern antiseptic medicine and hospital practice has been a major life saver for younger women during their childbearing years. But a look at the statistics reveals that childbirth fever was a minor cause of death during the nineteenth century, even of women of childbearing age, and was certainly not the cause of the excess mortality of women. Nearly all that excess mortality was a consequence of tuberculosis, and when tuberculosis ceased to be a major killer, women ceased to have a shorter life span than did men. A leading cause of mortality in young children was scalds and burns, especially among young girls because, of course, girls spent a great deal of time in very dangerous conditions, around open kitchen fires. Their young brothers spent a good deal of time outside the household, in workshops, admittedly not in the most favorable of working conditions, but somewhat less dangerous than the family hearth.

We return, then, to tuberculosis and the other infectious diseases that were such killers in the nineteenth century and the early part of the twentieth. An examination of the causes of death, first systematically recorded in the 1830s in Britain and a bit later in North America, show that most people did, indeed, die of infectious disease and in particular of respiratory diseases. They died of tuberculosis, of diphtheria, of bronchitis, of pneumonia, and particularly among children they died of measles and the perennial killer, smallpox. As the nineteenth century progressed, the death rate from all these diseases decreased continuously. Smallpox was dealt with by a medical advance, but one that could hardly be claimed by modern scientific medicine, since smallpox vaccine was discovered in the eighteenth century and already was quite widely used by the early part of the nineteenth. The death rates from the major killers like bronchitis, pneumonia, and tuberculosis fell rather regularly during the nineteenth century, with no obvious cause. There was no observable effect on the death rate after the germ theory of disease was announced in 1876 by Robert Koch. The death rate from these infectious diseases simply continued to decline as if Koch had never lived. By the time chemical therapy was introduced for tuberculosis in the earlier part of this century, more than 90 percent of the decrease in the death rate from that disease had already occurred.

One of the most revealing cases is measles. At present, Canadian and American children do not often get measles because they are vaccinated against it, but a generation ago every schoolchild had measles, yet death from measles was extremely rare. In the nineteenth century, measles was the major killer of young children, and in many African countries today it remains the highest cause of death among children. Measles is a disease that everyone used to contract, for which there is no known cure or medical treatment, and which simply stopped being fatal to children in advanced countries.

The progressive reductions in the death rate were not a consequence, for example, of modern sanitation, because the diseases that were the major killers in the nineteenth century were respiratory and not waterborne. It is unclear whether simple crowding had much to do with the process, since some parts of our cities are quite as crowded as they were in the 1850s. As far as we can tell, the decrease in death rates from the infectious killers of the nineteenth century is a consequence of the general improvement in nutrition and is related to an increase in the real wage. In countries like Brazil today, infant mortality rises and falls with decreases and increases in the minimum wage. The immense betterment of nutrition also explains the drop in the higher rate of tuberculosis among women than among men. In the nineteenth century, and even long into the twentieth in Britain, working men were far better nourished than home-bound women. Often if meat could be afforded for the table in an urban working-class family in Britain, it was saved for the man. So there have been complex social changes, resulting in increases in the real earnings of the great mass of people, reflected in part in their far better nutrition, that really lie at the basis of our increased longevity and our decreased death rate from infectious disease. Although one may say that the tubercle bacillus causes tuberculosis, we are much closer to the truth when we say that it was the conditions of unregulated nineteenth-century competitive capitalism, unmodulated by the demands of labor unions and the state, that was the cause of tuberculosis. But social causes are not in the ambit of biological science, so medical students continue to be taught that the cause of tuberculosis is a bacillus.

In the past 20 years, precisely because of the decline in infectious disease as an important cause of ill health, other single causes have been raised as the culprits in disease. It is undoubtedly true that pollutants and industrial wastes are the immediate physiological causes of cancers, miners' black lung, textile workers' brown lung, and a host of other disorders. Moreover, it is undoubtedly true that there are trace amounts of cancer-causing substances even in the best of our food and water unpolluted by pesticides and herbicides that make farm workers sick. But to say that pesticides cause the death of farm workers or that cotton fibers cause brown lung in textile workers is to make a fetish out of inanimate objects. We must distinguish between *agents* and *causes*. Asbestos fibers and pesticides are the agents of disease and disability, but it is illusory to suppose that if we eliminate these particular irritants that the diseases will go away, for other similar irritants will take their place. So long as efficiency, the maximization of profit from production, or the filling of centrally planned norms of production without reference to the means remain the motivating forces of productive enterprises the world over, so long as people are trapped by economic need or state regulation into production and consumption of certain things, then one pollutant will replace another. Regulatory agencies or central planning departments will calculate cost and benefit ratios where human misery is costed out at a dollar value. Asbestos and cotton lint fibers are not the causes of cancer. They are the agents of social causes, of social formations that determine the nature of our productive and consumptive lives, and in the end it is only through changes in those social forces that we can get to the root of problems of health. The transfer of causal power from social relations into inanimate agents that then seem to have a power and life of their own is one of the major mystifications of science and its ideologies.

Just as pollution is the most modern and up-to-date version of the external hostile forces of the physical world that are said to confront us, so simple internal forces, the genes, are now held responsible not only for human health in its normal medical sense but for a variety of social problems, among them alcoholism, criminality, drug addiction, and mental disorders. We are assured that if we could only find those genes that underlie alcoholism or the genes that have gone awry when we get cancer, then our problems will be over. The current manifestation of that belief in the importance of our inheritance in determining health and disease is the human genome sequencing project, a multibillion-dollar program of American and European biologists that is meant to take the place of space programs as the current great consumer of public money in the interest of conquering nature.

We know a great deal about what genes are made of and how they work at the most basic level. A gene is a long sequence of elements called nucleotides, of which there are only four kinds, identified by the letters A, T, C, and G. Every gene is a long string, of sometimes thousands or even tens of thousands of these A's, T's, C's, and G's, in a particular order: AATCCGGCATT and so on. This long sequence serves two functions. First, part of it specifies, like a code, exactly what the constitution of the protein molecules of our body will be. These proteins comprise the structural elements of which our bodies are made, the materials of our cells and tissues, and also the enzymes and hormones that make our metabolism possible. Corresponding to a particu-

lar sequence of A's, T's, C's, and G's, there will be produced by
the machinery of the body a long molecule, a protein made up of
simple elements, the amino acids. Each gene specifies the molecular makeup of a different protein. The particular sequence of
amino acids that constitutes a particular protein is determined by
the sequence of the nucleotides in the gene. If one or more
nucleotides in the gene are changed, a different amino acid may
be specified in the protein, which then may not be able to carry
on its physiological function as well as before. In some cases,
when a different nucleotide is substituted in a gene, less or even
none at all of a particular protein may be manufactured because
the machinery of the cell has a hard time recognizing the code.

Second, other parts of the gene, also sequences of nucleotides,
form part of the machinery that turns off and turns on the production of proteins. In this way, although the same genes are in
every part of the body during every part of the life of an organism,
proteins corresponding to some genes will be produced at some
times and in some parts of the body whereas they will not be
produced at other times and in other parts of the body. The
turning off and on of the production of the body's constituents is
itself sensitive to external conditions. For example, if the sugar
lactose is provided to the coliform bacterium, the presence of
the sugar will signal the bacterial machinery to start making a
protein that will break down the lactose and use it as a source of
energy. The signal to start translating the gene code into protein
is, in fact, detected by part of the gene itself. So, nucleotide
sequences determine what kind of proteins organisms will
make, and they are also part of the signaling machinery that
controls the manufacturing of those proteins in response to
external conditions. The signaling system is a mechanism by
which environment interacts with genes in creating organisms.

Genes have yet a further function, which is to serve as a
pattern for the manufacture of further copies of themselves.
When cells divide and sperm and egg cells are produced, every
new cell has a complete set of genes that are more or less
identical to the genes in the old cells. These newly manufactured

genes are copied directly from the gene molecules that previously existed. Since no chemical copying process is perfect, mistakes are made, so-called mutations, but these happen about one in a million copies as a rule.

The description I have just given of genes as determining the particular proteins that an organism can manufacture, as being part of the signaling system that responds to the environment in turning on and off the manufacture of protein, and as being the model for the manufacture of more of themselves, differs in a subtle way from the usual description of these relations. It is usually said that genes *make* proteins and that genes are *self-replicating*. But genes can *make* nothing. A protein is made by a complex system of chemical production involving other proteins, using the particular sequence of nucleotides in a gene to determine the exact formula for the protein being manufactured. Sometimes the gene is said to be the "blueprint" for a protein or the source of "information" for determining a protein. As such, it is seen as more important than the mere manufacturing machinery. Yet proteins cannot be manufactured without *both* the gene and the rest of the machinery. Neither is more important. Isolating the gene as the "master molecule" is another unconscious ideological commitment, one that places brains above brawn, mental work as superior to mere physical work, information as higher than action.

Nor are genes self-replicating. They cannot make themselves any more than they can make a protein. Genes are made by a complex machinery of proteins that uses the genes as models for more genes. When we refer to genes as self-replicating, we endow them with a mysterious, autonomous power that seems to place them above the more ordinary materials of the body. Yet if anything in the world can be said to be self-replicating, it is not the gene, but the entire organism as a complex system.

The human genome sequencing project is an ambitious plan to write down the complex nucleotide sequence of A's, T's, C's, and G's for all the genes of human beings. With current technology, this is an immensely ambitious project that might take 30 years and occupy tens or even hundreds of billions of dollars. Of course, there is always the promise that more efficient technology will become available to reduce the magnitude of the job. But why would one like to know the complete sequence of A's, T's, C's, and G's that make up all the human genes?

The claim is that if we had a reference sequence from a so-called normal individual and we compared bits and pieces of the sequence from a person with some disorder, then we could locate the genetic defect that causes the disease. We could then translate the genetic code of the altered person into an altered protein to see what is wrong with the protein, and this might tell us how to treat the disease. So, if diseases are caused by altered defective genes and if we know what a normal gene looks like in its finest molecular detail, we then would know what to do about fixing the abnormal physiology. We would know what proteins have gone awry in cancer and could somehow or another invent ways to fix them. We might find particular altered proteins or missing proteins in schizophrenics or in manic-depressives or in alcoholics or in drug-dependent people, and with an appropriate medication relieve them of these terrible disabilities. Moreover, by comparing all the human genes in their molecular detail with the genes of, say, a chimpanzee or a gorilla, we would know why we are different from chimpanzees. That is, we would know what it is to be human.

What is wrong with this vision? The first error it makes is in talking about the human gene sequence as if all human beings were alike. In fact, there is an immense amount of variation from normal individual to normal individual in the amino acid sequence of their proteins because a given protein may have a variety of amino acid compositions without impairing its function. Each of us carries two genes for each protein, one that we got from our mother and one from our father. On the average, the amino acid sequence specified by our maternally inherited and paternally inherited genes differ in about one every 12 genes. In addition, because of the nature of the genetic code, many changes occur at the level of DNA that are not reflected in proteins themselves. That is, there are many different DNA sequences that correspond to the same protein. We do not have good estimates for humans at the moment, but if humans are anything like experimental animals, about one in every 500 nucleotides will differ in DNA taken from any two individuals chosen at random. Since there are roughly 3 billion nucleotides in human genes, any two human beings will differ on the average in about 600,000 nucleotides. And an average gene that is, say, 3,000 nucleotides long will differ between any two normal individuals by about 20 nucleotides. Who's genome, then, is going to provide the sequence for the catalog for the normal person?

Moreover, every normal person carries a large number of defective genes in single dose inherited from one parent that are covered up by a normal copy that they received from their other parent. So any piece of DNA that is sequenced will have a certain number of unknown defective genes entered into the catalog. When the DNA from a person with a disease is compared to the DNA from the standard normal sequence, it would be impossible to decide which if any of the multiple differences between the two DNAs is responsible for the disease. It would be necessary to look at a large population of normal and diseased people to see if one could find some common difference between them, but even this may not happen if the disease in question has a multiple genetic cause so that different people have the same disease for different reasons, even if all those reasons are a consequence of genetic changes. We already know this to be the case for one human disease called thalassemia. Thalassemia is a blood disorder in which less than the normal amount of hemoglobin is made, a disorder many Asians and Mediterranean Europeans suffer. The deficiency is a consequence of defects in the gene that codes for the hemoglobin protein. It turns out there are at least 17 different defects in different parts of the hemoglobin gene, all of which result in a reduction in the amount of hemoglobin produced. We would look in vain for a particular nucleotide that differed between thalassemia and normal people. In the case of thalassemia, extensive population studies have revealed the correct story, but the possession of a standard normal sequence of the entire human genome was of no help here, nor would it be in any other case.

The second problem of the human genome sequencing project is that it also claims that in knowing the molecular configuration of our genes, we know everything that is worth knowing about us. It regards the gene as determining the individual, and the individual as determining society. It isolates an alteration in a so-called cancer gene as *the* cause of cancer, whereas that alteration in the gene may in turn have been caused by ingesting a pollutant, which in turn was produced by an industrial process, which in turn was the inevitable consequence of investing money at 6 percent. Once again, the impoverished notion of causation that characterizes modern biological ideology, a notion that confuses agents with causes, drives us in particular directions to find solutions for our problems.

Why, then, do so many powerful, famous, successful, and extremely intelligent scientists want to sequence the human genome? The answer is, in part, that they are so completely devoted to the ideology of simple unitary causes that they believe in the efficacy of the research and do not ask themselves more complicated questions. But in part the answer is a rather crass one. The participation in and the control of a multibillion-dollar, 30- or 50-year research project that will involve the everyday work of thousands of technicians and lower-level scientists is an extraordinarily appealing prospect for an ambitious biologist. Great careers will be made. Nobel Prizes will be given. Honorary degrees will be offered. Important professorships and huge laboratory facilities will be put at the disposal of those who control this project and who succeed in producing thousands of computer discs of human genome sequence. The realization that there are fairly straightforward economic and status rewards awaiting those who take part in the project has given rise to a powerful opposition to the project from within biology itself, from others who are doing a different kind of science and see their own careers and their own research threatened by the diversion of money, energy, and public consciousness. Some farsighted biologists have cautioned against the terrible public disillusionment that will follow the completion of the human genome sequencing project. The public will discover that despite the inflated claims of molecular biologists, people are still dying of cancer, of heart disease, of stroke, that institutions are still filled with schizophrenics and manic-depressives, that the war against drugs has not been won. The fear among many scientists is that by promising too much, science will destroy its public image and people will become cynical as, for example, they became cynical about the war on cancer, not to speak of the war on poverty.

Research scientists are involved in this struggle not only in their role as academics. Among molecular biologists who are professors in universities, a large proportion are also principal scientists or principal stockholders in biotechnology companies. Technology is a major industry and a major source of hoped-for profit for venture capital. The human genome sequencing project, to the extent that it creates new technologies at public expense, will provide very powerful tools to biotechnology companies for carrying out their production of commodities for sale on the market. Moreover, the success of the project will give greater faith in the power of biotechnology to produce useful products.

Nor are biotechnologies the only producers of commodities that stand to gain immensely from the human genome sequencing project. The project will consume vast quantities of chemical and mechanical commodities. There are commercial machines that manufacture DNA from small amounts of sample material. There are machines that automatically sequence DNA. These machines require the input of a variety of chemical materials all of which are sold at an immense profit by the companies that manufacture the machines. The human genome sequencing project is big business. The billions of dollars that are to be spent on it will go in no insignificant fraction into the annual dividends of productive enterprises.

We see in the genome sequencing project an aspect of biological science that is not often spoken of and is perhaps the most mystified of all. What are said to be fundamental discoveries about the nature of life often mask simple commercial relations that provide a powerful impetus for the direction and subject of research. The best documented example that we have of purely commercial interest driving what is said to be a fundamental discovery about nature is in agriculture. It is commonly said that the invention of hybrid corn has resulted in immense increases in the productivity of agriculture and the consequent feeding of hundreds of millions of people at low cost and with great efficiency. Whereas in the 1920s an average acre sown with maize in the cornbelt of North America might have yielded 35 bushels per acre, today it can yield 125 bushels. This is widely regarded as one of the greatest triumphs of basic genetics as applied to human welfare. But the truth is more interesting.

Hybrid corn is produced by crossing two true-breeding inbred varieties and planting the seed from that cross. These true-breeding inbred varieties are created by a long process of self-pollination to make each variety completely uniform genetically. A seed company will spend a certain number of years self-pollinating lines of corn until it gets uniform lines, and it will then sell to the farmer the seed that comes from crossing two of these lines. The inbred homogeneous lines themselves give rather poor yields, whereas the hybrid is superior in productivity both to the inbred lines and to the original open-pollinated population of corn from which the inbreds came. It is not the case that a cross between any two inbred homogeneous lines of corn will produce a hybrid with high yield. It is necessary to search among many such homogeneous inbred lines to find pairs that will do the trick.

The hybrid cross between the inbred lines has another quality, which is not much spoken about, a quality with a unique commercial value. If a farmer has a high-yielding variety of some crop, one that is resistant to disease and produces high commercial output as compared to the cost of inputs, his normal way of carrying on his business would be to save some of the seed of this high-yielding variety and plant it next year to again achieve high yields. Once the farmer acquired the seed of this wonderful variety, he would no longer have to pay again to reacquire it, because plants, like other organisms, reproduce themselves. But this self-reproduction presents a serious problem to someone who wants to make money by developing new varieties of organisms. For how will he make a profit if the moment he has sold the seed, its further production is in the hands of the person who bought it? He will get to sell it one time only, and then it will be distributed everywhere for nothing.

This is the problem of copy protection that also exists for computer software programs. The developer of computer software will be unwilling to devote time, energy, and money to developing a new program if the first purchasers can copy it and pass it around to their friends for virtually nothing. That has always been the problem of plant and animal breeding. Plant breeders and seed producers could never make much money because farmers, having bought the seed or the animal variety, would produce future generations themselves. Of course, seeds produced on the farm may contain a certain amount of weed seed and not be produced under the best conditions for germination so, in fact, farmers will occasionally go back to the seed producer for a new stock. In France, for example, the average wheat farmer goes back once every six years to replenish the supply of wheat seed.

Hybrid corn is different. Because it is the cross between two self-propagating homogeneous lines, one cannot plant the seed of hybrid corn and get new hybrid corn. Hybrids are not true-breeding. The seeds that are borne on a hybrid plant are not themselves hybrids but form a population of plants of varying degrees of hybridity, a mixture of homogeneous and heterogeneous varieties. A farmer who saved seed from his hybrid corn and planted it the next year would lose at least 30 bushels per acre in the next crop. To maintain high yields, it is necessary for the farmer to go back every year and buy the seed again. So, the hybrid seed corn producer has found a method of copy protection. Moreover, the producer of the hybrid seed can charge the farmer a price for the hybrid seed that is equivalent to the amount the farmer would have lost—that is, the market value of 30 bushels per acre—had he not returned to the seed company for more hybrid seed. The invention of hybrid corn was, in fact, a deliberate use of the principles of genetics to create a copy-protected product. We have that on the best authority possible, the inventors of hybrid corn themselves, Shull and East, who wrote that hybrids are

> something that might easily be taken up by the seedsmen; in fact, it is the first time in agricultural history that a seedsman is enabled to gain the full benefit from a desirable origination of his own or something that he has purchased ... The man who originates a new plant which may be of incalculable benefit to the whole country gets nothing—not even fame— for his pains and the plant can be propagated by anyone ... The utilization of the first generation hybrids enables the originator to keep the parental types and give out only the crossed seeds, which are less valuable for continued propagation.[1]

The realization that the hybrid method could guarantee to the inventor immense profits has resulted in the introduction of the hybrid method into all of agriculture. Chickens, tomatoes, swine, indeed, nearly every commercial plant or animal where it is possible to introduce the method has seen the growth of hybrids at the expense of older varietal forms. Major seed companies, like the Pioneer Hybrid Seed Company, have invested millions of dollars in attempting to produce hybrid wheat that would then capture an immense untapped market. So far, they have not succeeded, because the cost of production of the hybrid seed is excessive.

The Pioneer Hybrid Seed Company itself is the consequence of the activities of a single important political and scientific figure, Henry A. Wallace. Wallace's father was appointed secretary of agriculture of the United States by President Warren

Harding in 1920. The elder Wallace sent Henry on a tour of agricultural experiment stations. On his return, Henry advised his father to appoint as head of plant breeding a man who was devoted to hybrids. In the meanwhile, Henry was himself experimenting with hybrids, and in 1924 he sold his first hybrid seed corn at a profit of about $740 an acre. In 1926, he founded the Pioneer Hybrid Seed Company, and when, in 1932, he was appointed secretary of agriculture by President Franklin Roosevelt, pressure for the introduction of hybrid corn in the United States, and subsequently in Canada, became irresistible.

If hybrids really are a superior method for agricultural production, then their commercial usefulness to the seed company is a side issue. The question is whether other methods of plant breeding might have worked as well or better without providing property-rights protection for the seed companies. The answer to that question depends on some issues in basic genetics that were undecided in the early history of hybrid corn, and until 30 years ago, one might have argued that the basic biology of corn production is such that only hybrids would provide the added yield. However, we have known the truth of the matter for the last 30 years. The fundamental experiments have been done and no plant breeder disagrees with them. The nature of the genes responsible for influencing corn yield is such that the alternative method of simple direct selection of high-yielding plants in each generation and the propagation of seed from those selected plants would work. By the method of selection, plant breeders could, in fact, produce varieties of corn that yield quite as much as modern hybrids. The problem is that no commercial plant breeder will undertake such investigation and development because there is no money in it.

One of the most interesting features of this story is the role of agricultural experiment stations like the state agricultural experiment stations in the United States or the Canadian Department of Agriculture. These institutions might be expected to develop alternative methods since they are not concerned with profit and are working at public expense. Yet the U.S. Department of

Agriculture and the Canadian Department of Agriculture are among the strongest proponents of the hybrid method. A purely commercial interest has so successfully clothed itself in the claims of pure science that those claims are now taught as scientific gospel in university schools of agriculture. Successive generations of agricultural research workers, even those who work in public institutions, believe that hybrids are intrinsically superior even though the experimental results that contradict this have been published in well-known journals for more than 30 years. Once again, what appears to us in the mystical guise of pure science and objective knowledge about nature turns out, underneath, to be political, economic, and social ideology.

THE DREAM OF
THE HUMAN
GENOME

The practical outcome of the belief that what we want to know about human beings is contained in the sequence of their DNA is the Human Genome Project in the United States and, in its international analogue, the Human Genome Organization (HUGO), called by one molecular biologist "the UN for the human genome."

These projects are, in fact, administrative and financial organizations rather than research projects in the usual sense. They have been created over the last five years in response to an active lobbying effort by scientists such as Walter Gilbert, James Watson, Charles Cantor, and Leroy Hood, aimed at capturing very large amounts of public funds and directing the flow of those funds into an immense cooperative research program.

The ultimate purpose of this program is to write down the complete ordered sequence of A's, T's, C's, and G's that make up all the genes in the human genome, a string of letters that will be 3 billion elements long. The first laborious technique for cutting up DNA nucleotide by nucleotide and identifying each nucleotide in order as it is broken off was invented fifteen years ago by Allan Maxam and Walter Gilbert, but since then the process has become mechanized. DNA can now be squirted into one end of a mechanical process and out the other end will emerge a four-color computer printout announcing "AGGACTT. . . ." In the course of the genome project yet more efficient mechanical schemes will be invented and complex computer programs will be developed to catalogue, store, compare, order, retrieve, and otherwise organize and reorganize the immensely long string of letters that will emerge from the machine. The work will be a collective enterprise of very large laboratories, "Genome Centers," that are to be specially funded for the purpose.

The project is to proceed in two stages. The first is so-called "physical mapping." The entire DNA of an organism is not one

This chapter was not part of the original Massey Lectures and is reprinted from *The New York Review of Books*, vol. 39, no. 10, May 28, 1992.

long unbroken string, but is divided up into a small number of units, each of which is contained in one of a set of microscopic bodies in the cell, the chromosomes. Human DNA is broken up into twenty-three different chromosomes, while fruit flies' DNA is contained in only four chromosomes. The mapping phase of the genome project will determine short stretches of DNA sequence spread out along each chromosome as positional landmarks, much as mile markers are placed along superhighways. These positional markers will be of great use in finding where in each chromosome particular genes may lie. In the second phase of the project, each laboratory will take a chromosome or a section of a chromosome and determine the complete ordered sequence of nucleotides in its DNA. It is after the second phase, when the genome project, *sensu strictu*, has ended, that the fun begins, for biological sense will have to be made, if possible, of the mind-numbing sequence of three billion A's, T's, C's, and G's. What will it tell us about health and disease, happiness and misery, the meaning of human existence?

The American project is run jointly by the National Institutes of Health and the Department of Energy in a political compromise over who should have control over the hundreds of millions of dollars of public money that will be required. The project produces a glossy-paper newsletter distributed free, headed by a coat of arms showing a human body wrapped Laocoön-like in the serpent coils of DNA and surrounded by the motto, "Engineering, Chemistry, Biology, Physics, Mathematics." The Genome Project is the nexus of all sciences. My latest copy of the newsletter advertises the free loan of a twenty-three-minute video on the project "intended for high school age and older," featuring, among others, several of the contributors to a collection of essays about the Human Genome Project, *The Code of Codes*, and a calendar of fifty "Genome Events."

Another consequence of the conviction that DNA contains the secret of human life is the appearance of a large number of popular and semi-popular books touting the wonders of the Human Genome Project.[1] None of the authors of these books seems to be in any doubt about the importance of the project to determine the complete DNA sequence of a human being. "The Most Astonishing Adventure of Our Time," say Jerry E. Bishop and Michael Waldholz; "The Future of Medicine," according to Lois Wingerson; "today's most important scientific undertaking," dictating "The Choices of Modern Science," Joel Davis declares in *Mapping the Code*.

Nor are these simply the enthusiasms of journalists. The molecular biologist Christopher Wills says that "the outstanding problems in human biology . . . will all be illuminated in a strong and steady light by the results of this undertaking"; the great panjandrum of DNA himself, James Dewey Watson, explains, in his essay in the collection edited by Kevles and Hood, that he doesn't "want to miss out on learning how life works," and Gilbert predicts that there will be "a change in our philosophical understanding of ourselves." Surely, "learning how life works" and "a change in our philosophical understanding of ourselves" must be worth a lot of time and money. Indeed, there are said to be those who have exchanged something a good deal more precious for that knowledge.

Unfortunately, it takes more than DNA to make a living organism. I once heard one of the world's leaders in molecular biology say, in the opening address of a scientific congress, that if he had a large enough computer and the complete DNA sequence of an organism, he could compute the organism, by which he meant totally describe its anatomy, physiology, and behavior. But that is wrong. Even the organism does not compute itself from its DNA. A living organism at any moment in its life is the unique consequence of a developmental history that results from the interaction of and determination by internal and external forces. The external forces, what we usually think of as "environment," are themselves partly a consequence of the activities of the organism itself as it produces and consumes the conditions of its own existence. Organisms do not find the world in which they develop. They make it. Reciprocally, the internal forces are not autonomous, but act in response to the external. Part of the internal chemical machinery of a cell is only manufactured when external conditions demand it. For example, the enzyme that breaks down the sugar, lactose, to provide energy for bacterial growth is only manufactured by bacterial cells when they detect the presence of lactose in their environment.

Nor is "internal" identical with "genetic." Fruit flies have long hairs that serve as sensory organs, rather like a cat's whiskers. The number and placement of those hairs differ between the two sides of a fly (as they do between the left and right sides of a cat's muzzle), but not in any systematic way. Some flies have more hairs on the left, some more on the right. Moreover, the variation between sides of a fly is as great as the average variation from fly to fly. But the two sides of a fly have the same genes and have had the same environment during development. The variation between sides is a consequence of random cellular movements and chance molecular events within cells during development, so-called "developmental noise." It is this same developmental noise that accounts for the fact that identical twins have different fingerprints and that the fingerprints on our left and right hands are different. A desktop computer that was as sensitive to room temperature and as noisy in its internal circuitry as a developing organism could hardly be said to compute at all.

The scientists writing about the Genome Project explicitly reject an absolute genetic determinism, but they seem to be writing more to acknowledge theoretical possibilities than out of conviction. If we take seriously the proposition that the internal and external codetermine the organism, we cannot really believe that the sequence of the human genome is the grail that will reveal to us what it is to be human, that it will change our philosophical view of ourselves, that it will show how life works. It is only the social scientists and social critics, such as Kevles, who comes to the Genome Project from his important study of the continuity of eugenics with modern medical genetics; Dorothy Nelkin, both in her book with Laurence Tancredi and in her chapter in Kevles and Hood; and, most strikingly, Evelyn Fox Keller in her contribution to *The Code of Codes*, for whom the problem of the development of the organism is central.

Nelkin, Tancredi, and Keller suggest that the importance of the Human Genome Project lies less in what it may, in fact, reveal about biology, and whether it may in the end lead to a successful therapeutic program for one or another illness, than in its validation and reinforcement of biological determinism as an explanation of all social and individual variation. The medical model that begins, for example, with a genetic explanation of the extensive and irreversible degeneration of the central nervous system characteristic of Huntington's chorea, may end with an explanation of human intelligence, of how much people drink, how intolerable they find the social condition of their lives, whom they choose as sexual partners, and whether they get sick on the job. A medical model of all human variation makes a medical model of normality, including social normality, and dictates that we preemptively or through subsequent corrective therapy bring into line anyone who deviates from that norm.

There are many human conditions that are clearly pathological and that can be said to have a unitary genetic cause. As far as is known, cystic fibrosis and Huntington's chorea occur in people carrying the relevant mutant gene irrespective of diet, occupation, social class, or education. Such disorders are rare: 1 in 2,300 births for cystic fibrosis, 1 in 3,000 for Duchenne's muscular dystrophy, 1 in 10,000 for Huntington's disease. A few other conditions occur in much higher frequency in some populations, but are generally less severe in their effects and more sensitive to environmental conditions, as for example sickle-cell anemia in West Africans and their descendants, who suffer severe effects only in conditions of physical stress. These disorders provide the model on which the program of medical genetics is built, and they provide the human interest drama on which books like *Mapping our Genes* and *Genome* are built. In reading them, I saw again those heroes of my youth, Edward G. Robinson curing syphilis in *Dr. Ehrlich's Magic Bullet*, and Paul Muni saving children from rabies in *The Story of Louis Pasteur*.

According to the vision of the project and its disciples, we will locate on the human chromosomes all the defective genes that plague us, and then from the sequence of the DNA we will deduce the causal story of the disease and generate a therapy. Indeed, a great many defective genes have already been roughly mapped onto chromosomes and, with the use of molecular techniques, a few have been very closely located and, for even fewer, some DNA sequence information has been obtained. But causal stories are lacking and therapies do not yet exist; nor is it clear, when actual cases are considered, how therapies will flow from a knowledge of DNA sequences.

The gene whose mutant form leads to cystic fibrosis has been located, isolated, and sequenced. The protein encoded by the gene has been deduced. Unfortunately, it looks like a lot of other proteins that are a part of cell structure, so it is hard to know what to do next. The mutation leading to Tay-Sachs disease is even better understood because the enzyme specified by the gene has a quite specific and simple function, but no one has suggested a therapy. On the other hand, the gene mutation causing Huntington's disease has eluded exact location, and no biochemical or specific metabolic defect has been found for a disease that results in catastrophic degeneration of the central nervous system in every carrier of the defective gene.

A deep reason for the difficulty in devising causal information from DNA messages is that the same "words" have different meanings in different contexts and multiple functions in a given context, as in any complex language. No word in English has more powerful implications of action than "do." "Do it now!" Yet in most of its contexts "do" as in "I do not know" is periphrastic, and has no meaning at all. While the periphrastic "do" has no *meaning*, it undoubtedly has a linguistic *function* as a place holder and spacing element in the arrangement of a sentence. Otherwise, it would not have swept into general English usage in the sixteenth century from its Midlands dialect origin, replacing everywhere the older "I know not."

So elements in the genetic messages may have meaning, or

they may be periphrastic. The code sequence GTAAGT is sometimes read by the cell as an instruction to insert the amino acids *valine* and *serine* in a protein, but sometimes it signals a place where the cell machinery is to cut up and edit the message; and sometimes it may be only a spacer, like the periphrastic "do," that keeps other parts of the message an appropriate distance from each other. Unfortunately, we do not know how the cell decides among the possible interpretations. In working out the interpretive rules, it would certainly help to have very large numbers of different gene sequences, and I sometimes suspect that the claimed significance of the genome sequencing project for human health is an elaborate cover story for an interest in the hermeneutics of biological scripture.

Of course, it can be said, as Gilbert and Watson do in their essays, that an understanding of how the DNA code works is the path by which human health will be reached. If one had to depend on understanding, however, we would all be much sicker than we are. Once, when the eminent Kant scholar, Lewis Beck, was traveling in Italy with his wife, she contracted a maddening rash. The specialist they consulted said it would take him three weeks to find out what was wrong with her. After repeated insistence by the Becks that they had to leave Italy within two days, the physician threw up his hands and said, "Oh, very well, Madame. I will give up my scientific principles. I will cure you today."

Certainly an understanding of human anatomy and physiology has led to a medical practice vastly more effective than it was in the eighteenth century. These advances, however, consist in greatly improved methods for examining the state of our insides, of remarkable advances in microplumbing, and of pragmatically determined ways of correcting chemical imbalances and of killing bacterial invaders. None of these depends on a deep knowledge of cellular processes or on any discoveries of molecular biology. Cancer is still treated by gross physical and chemical assaults on the offending tissue. Cardiovascular disease is treated by surgery whose anatomical bases go back to the nineteenth century, by diet, and by pragmatic drug treatment. Antibiotics were originally developed without the slightest notion of how they do their work. Diabetics continue to take insulin, as they have for sixty years, despite all the research on the cellular basis of pancreatic malfunction. Of course, intimate knowledge of the living cell and of basic molecular processes may be useful eventually, and we are promised over and over that results are just around the corner. But as Miss Adelaide so poignantly complained:

> *You promise me this*
> *You promise me that.*
> *You promise me everything*
>   *under the sun.*

> *When I think of the time gone by*
> *I could honestly die.*

Not the least of the problems of turning sequence information into causal knowledge is the existence of large amounts of "polymorphism." While the talk in most of the books under review is of sequencing the human genome, every human genome differs from every other. The DNA I got from my mother differs by about one tenth of one percent, or about 3,000,000 nucleotides, from the DNA I got from my father, and I differ by about that much from any other human being. The final catalogue of "the" human DNA sequence will be a mosaic of some hypothetical average person corresponding to no one. This polymorphism has several serious consequences. First, all of us carry one copy, inherited from one parent, of mutations that would result in genetic diseases if we had inherited two copies. No one is free of these, so the catalogue of the standard human genome after it is compiled will contain, unknown to its producers, some fatally misspelled sequences which code for defective proteins or no protein at all. The only way to know whether the standard sequence is, by bad luck, the code of a defective gene is to sequence the same part of the genome from many different individuals. Such polymorphism studies are not part of the Human Genome Project and attempts to obtain money from the project for such studies have been rebuffed.

Second, even genetically "simple" diseases can be very heterogeneous in their origin. Sequencing studies of the gene that codes for a critical protein in blood-clotting has shown that hemophiliacs differ from people whose blood clots normally by any one of 208 different DNA variations, all in the same gene. These differences occur in every part of the gene, including bits that are not supposed to affect the structure of the protein.

The problem of telling a coherent causal story, and of then designing a therapy based on knowledge of the DNA sequence in such a case, is that we do not know even in principle all of the functions of the different nucleotides in a gene, or how the specific context in which a nucleotide appears may affect the way in which the cell machinery interprets the DNA; nor do we have any but the most rudimentary understanding of how a whole functioning organism is put together from its protein bits and pieces. Third, because there is no single, standard, "normal" DNA sequence that we all share, observed sequence differences between sick and well people cannot, in themselves, reveal the genetic cause of a disorder. At the least, we would need the sequences of many sick and many well people to look for common differences between sick and well. But if many diseases are like hemophilia, common differences will not be found and we will remain mystified.

The failure to turn knowledge into therapeutic power does not discourage the advocates of the Human Genome Project because their vision of therapy includes *gene* therapy. By techniques that are already available and need only technological development, it is possible to implant specific genes containing the correct gene sequence into individuals who carry a mutated sequence, and to induce the cell machinery of the recipient to use the implanted genes as its source of information. Indeed, the first case of human gene therapy for an immune disease—the treatment of a child who suffered from a rare disorder of the immune system— has already been announced and seems to have been a success. The supporters of the Genome Project agree that knowing the sequence of all human genes will make it possible to identify and isolate the DNA sequences for large numbers of human defects which could then be corrected by gene therapy. In this view, what is now an *ad hoc* attack on individual disorders can be turned into a routine therapeutic technique, treating every physical and psychic dislocation, since everything significant about human beings is specified by their genes.

However, gene implantation may affect not only the cells of our temporary bodies, our *somatic* cells, but the bodies of future generations through accidental changes in the *germ* cells of our reproductive organs. Even if it were our intention only to provide properly functioning genes to the immediate body of the sufferer, some of the implanted DNA might get into and transform future sperm and egg cells. Then future generations would also have undergone the therapy *in absentia* and any miscalculations of the effects of the implanted DNA would be wreaked on our descendants to the remotest time. So David Suzuki and Peter Knudtson make it one of their principles of "genethics" (they have self-consciously created ten of them) that

> while genetic manipulation of human somatic cells may lie
> in the realm of personal choice, tinkering with human germ
> cells does not. Germ-cell therapy, without the consent of all
> members of society, ought to be explicitly forbidden.

Their argument against gene therapy is a purely prudential one, resting on the imprecision of the technique and the possibility that a "bad" gene today might turn out to be useful someday. This seems a slim base for one of the Ten Commandments of biology, for, after all, the techniques may get a lot better and mistakes can always be corrected by another round of gene therapy. The vision of power offered to us by gene therapists makes gene transfer seem rather less permanent than a silicone implant or a tummy tuck. The bits of ethics in *Genethics* is, like a Unitarian sermon, nothing that any decent person could quarrel with. Most of the "genethic principles" turn out to be, in fact, prudential advice about why we should not screw around with our genes or those of other species. While most of their arguments are sketchy, Suzuki and Knudtson are the only authors among those under review who take seriously the problems presented by genetic diversity among individuals, and who attempt to give the reader enough understanding of the principles of population genetics to think about these problems.

Most death, disease, and suffering in rich countries do not arise from muscular dystrophy and Huntington's chorea, and, of course, the majority of the world's population is suffering from one consequence or another of malnutrition and overwork. For Americans, it is heart disease, cancer, and stroke that are the major killers, accounting for 70 percent of deaths, and about 60 million people suffer from chronic cardiovascular disease. Psychiatric suffering is harder to estimate, but before the psychiatric hospitals were emptied in the 1960s, there were 750,000 psychiatric inpatients. It is now generally accepted that some fraction of cancers arise on a background of genetic predisposition. That is, there are a number of genes known, the so-called *oncogenes*, that have information about normal cell division. Mutations in these genes result (in an unknown way) in making cell division less stable and more likely to occur at a pathologically high rate. Although a number of such genes have been located, their total number and the proportion of all cancers influenced by them is unknown.

In no sense of simple causation are mutations in these genes *the* cause of cancer, although they may be one of many predisposing conditions. Although a mutation leading to extremely elevated cholesterol levels is known, the great mass of cardiovascular disease has utterly defied genetic analysis. Even diabetes, which has long been known to run in families, has never been tied to genes and there is no better evidence for a genetic predisposition to it in 1992 than there was in 1952 when serious genetic studies began. No week passes without the announcement in the press of a "possible" genetic cause of some human ill which upon investigation "may eventually lead to a cure." No literate public is unassailed by the claims. The *Morgunbladid* of Reykjavik asks its readers rhetorically, *"Med allt í genunum?"* ("Is it all in the genes?") in a Sunday supplement.

The rage for genes reminds us of Tulipomania and the South Sea Bubble in McKay's *Great Popular Delusions and the Madness of Crowds*. Claims for the definitive location of a gene for schizophrenia and manic depressive syndrome using DNA markers have been followed repeatedly by retraction of the claims and contrary claims as a few more members of a family tree have been observed, or a different set of families examined. In one notorious case, a claimed gene for manic depression, for which there was strong statistical evidence, was nowhere to be found when two members of the same family group developed symptoms. The original claim and its retraction both were published in the international journal *Nature*, causing David Baltimore to cry out at a scientific meeting, "Setting myself up as an average reader of *Nature*, what am I to believe?" Nothing.

Some of the wonder-workers and their disciples see even beyond the major causes of death and disease. They have an image of social peace and order emerging from the DNA data bank at the National Institutes of Health. The editor of the most prestigious general American scientific journal, *Science*, an energetic publicist for large DNA sequencing projects, in special issues of his journal filled with full-page multicolored advertisements from biotechnology equipment manufacturers, has visions of genes for alcoholism, unemployment, domestic and social violence, and drug addiction. What we had previously imagined to be messy moral, political, and economic issues turn out, after all, to be simply a matter of an occasional nucleotide substitution. While the notion that the war on drugs will be won by genetic engineering belongs to Cloud Cuckoo Land, it is a manifestation of a serious ideology that is continuous with the eugenics of an earlier time.

Daniel Kevles has quite persuasively argued in his earlier book on eugenics that classical eugenics became transformed from a social program of general population improvement into a family program of providing genetic knowledge to individuals facing reproductive decisions. But the ideology of biological determinism on which eugenics was based has persisted and, as is made clear in Kevles's excellent short history of the Genome Project in *The Code of Codes*, eugenics in the social sense has been revivified. This has been in part a consequence of the mere existence of the Genome Project, with its accompanying public relations, and the heavy public expenditure it will require. These alone validate its determinist *Weltanschauung*. The publishers declare the glory of DNA and the media showeth forth its handiwork.

The nine books I have mentioned here are only a sample of what has been and what is to come. The cost of sequencing the human genome is estimated optimistically at 300 million dollars (10 cents a nucleotide for the 3 billion nucleotides of the entire genome), but if development costs are included it surely cannot be less than half a billion in current dollars. In fact, the managers of the project are hoping for a budget of 200 million dollars a year for fifteen years. Moreover the genome project *sensu strictu* is only the beginning of wisdom. Yet more hundreds of millions must be spent on chasing down the elusive differences in DNA for each specific genetic disease, of which some 3,000 are now known, and some considerable fraction of that money will stick to entrepreneurial molecular geneticists. None of our authors has the bad taste to mention that many molecular geneticists of repute, including several of the essayists in *The Code of Codes*, are founders, directors, officers, and stockholders in commercial biotechnology firms, including the manufacturers of the supplies and equipment used in sequencing research. Not all authors have Norman Mailer's openness when they write advertisements for themselves.

It has been clear since the first discoveries in molecular biology that "genetic engineering," the creation to order of genetically altered organisms, has an immense possibility for producing private profit. If the genes that allow clover plants to manufacture their own fertilizer out of the nitrogen in the air could be transferred to maize or wheat, farmers would save great sums and the producers of the engineered seed would make a great deal of money. Genetically engineered bacteria grown in large fermenting vats can be made into living factories to produce rare and costly molecules for the treatment of viral diseases and cancer. A bacterium has already been produced that will eat raw petroleum, making oil spills biodegradable. As a consequence of these possibilities, molecular biologists have become entrepreneurs. Many have founded biotechnology firms funded by venture capitalists. Some have become very rich when a successful public offering of their stock has made them suddenly the holders of a lot of valuable paper. Others find themselves with large blocks of stock in international pharmaceutical companies who have bought out the biologist's mom-and-pop enterprise and acquired their expertise in the bargain.

No prominent molecular biologist of my acquaintance is without a financial stake in the biotechnology business. As a result, serious conflicts of interest have emerged in universities and in government service. In some cases graduate students working under entrepreneurial professors are restricted in their scientific interchanges, in case they might give away potential trade secrets. Research biologists have attempted, sometimes with success, to get special dispensations of space and other resources from their universities in exchange for a piece of the action. Biotechnology joins basketball as an important source of educational cash.

Public policy too, reflects private interest. James Dewey Watson resigned in April as head of the NIH Human Genome Office as a result of pressure put on him by Bernardine Healey, Director of the NIH. The immediate form of this pressure was an investigation by Healey of the financial holdings of Watson or his immediate family in various biotechnology firms. But nobody in the molecular biological community believes in the seriousness of such an investigation, because everyone including Dr. Healey knows that there are no financially disinterested candidates for Watson's job. What is really at issue is a disagreement about patenting the human genome. Patent law prohibits the patenting of anything that is "natural," so, for example, if a rare plant were discovered in the Amazon whose leaves could cure cancer, no one could patent it. But, it is argued, isolated genes are not natural, even though the organism from which they are taken may be. If human DNA sequences are to be the basis of future therapy, then the exclusive ownership of such DNA sequences would be money in the bank.

Dr. Healey wants the NIH to patent the human genome to prevent private entrepreneurs, and especially foreign capital, from controlling what has been created with American public funding. Watson, whose family is reported to have a financial stake in the British pharmaceutical firm Glaxo, has characterized Healey's plan as "sheer lunacy," on the grounds that it will slow down the acquisition of sequence information. (Watson has denied any conflict of interest.) Sir Walter Bodmer, the director of the Imperial Cancer Research Fund, and a major figure in the European Genome organization, spoke the truth that we all know lies behind the hype of the Human Genome Project when he told *The Wall Street Journal* that "the issue [of ownership] is at the heart of everything we do."

The study of DNA is an industry with high visibility, a claim on the public purse, the legitimacy of a science, and the appeal that it will alleviate individual and social suffering. So its basic ontological claim, of the dominance of the Master Molecule over the body physical and the body politic, becomes part of general consciousness. Evelyn Fox Keller's chapter in *The Code of Codes* brilliantly traces the percolation of this consciousness through the strata of the state, the universities, and the media, producing an unquestioned consensus that the model of cystic fibrosis is a model of the world. Daniel Koshland, the editor of *Science*, when asked why the Human Genome Project funds should not be given instead to the homeless, answered, "What these people don't realize is that the homeless are impaired. . . . Indeed, no group will benefit more from the application of human genetics."[2]

Beyond the building of a determinist ideology, the concentration of knowledge about DNA has direct practical, social and political consequences, what Dorothy Nelkin and Laurence Tancredi call "The Social Power of Biological Information." Intellectuals in their self-flattering wish-fulfillment say that knowledge is power, but the truth is that knowledge further empowers only those who have or can acquire the power to use it. My possession of a Ph.D. in nuclear engineering and the complete plans of a nuclear power station will not reduce my electric bill by a penny. So with the information contained in DNA, there is no instance where knowledge of one's genes does not further concentrate the existing relations of power between individuals and between the individual and institutions.

When a woman is told that the fetus she is carrying has a 50 percent chance of contracting cystic fibrosis, or for that matter that it will be a girl although her husband desperately wants a boy, she does not gain additional power just by having that knowledge, but is only forced by it to decide and to act within the confines of her relation to the state and her family. Will her husband agree to or demand, an abortion, will the state pay for it, will her doctor perform it? The slogan "a woman's right to choose" is a slogan about conflicting relations of power, as Ruth Schwartz Cowan makes clear in her essay "Genetic Technology and Reproductive Choice: An Ethics for Autonomy" in *The Code of Codes*.

Increasingly, knowledge about the genome is becoming an element in the relation between individuals and institutions, generally adding to the power of institutions over individuals. The relations of individuals to the providers of health care, to the schools, to the courts, to employers are all affected by knowledge, or the demand for knowledge, about the state of one's DNA. In the essays by both Henry Greeley and Dorothy Nelkin in *The Code of Codes*, and in much greater detail and extension in *Dangerous Diagnostics*, the struggle over biological information is revealed. The demand by employers for diagnostic information about the DNA of prospective employees serves the firm in two ways. First, as providers of health insurance, either directly or through their payment of premiums to insurance companies, employers reduce their wage bill by hiring only workers with the best health prognoses. Second, if there are workplace hazards to which employees may be in different degrees sensitive, the employer may refuse to employ those whom it judges to be sensitive. Not only does such employment exclusion reduce the potential costs of health insurance, but it shifts the responsibility of providing a safe and healthy workplace from the employer to the worker. It becomes the worker's responsibility to look for work that is not threatening. After all, the employer is helping the workers by providing a free test of susceptibilities and so allowing them to make more informed choices of the work they would like to do. Whether other work is available at all, or worse paid, or more dangerous in other ways, or only in a distant place, or extremely unpleasant and debilitating is simply part of the conditions of the labor market. So Koshland is right after all. Unemployment and homelessness do indeed reside in the genes.

Biological information has also become critical in the relation between individuals and the state, for DNA has the power to put a tongue in every wound. Criminal prosecutors have long hoped for a way to link accused persons to the scene of a crime when there are no fingerprints. By using DNA from a murder victim and comparing it with DNA from dried blood found on the person or property of the accused, or by comparing the accused's DNA with DNA from skin scrapings under the fingernails of a rape victim, prosecutors attempt to link criminal and crime. Because of the polymorphism of DNA from individual to individual, a definitive identification is, in principle, possible. But, in practice, only a bit of DNA can be used for identification so there is some chance that the accused will match the DNA from the crime scene even though someone else is in fact guilty.

Moreover, the methods used are prone to error, and false matches (as well as false exclusions) can occur. For example, the FBI characterized the DNA of a sample of 225 FBI agents and then, on a retest of the same agents, found a large number of mismatches. Matching is almost always done at the request of the prosecutor, because tests are expensive and most defendants in assault cases are represented by a public defender or court-appointed lawyer. The companies who do the testing have a vested commercial interest in providing matches, and the FBI, which also does some testing, is an interested party.

Because different ethnic groups differ in the frequency of the various DNA patterns, there is also the problem of the appropriate reference group to whom the defendant is to be compared. The identity of that reference group depends in complex ways on the circumstances of the case. If a woman who is assaulted lives in Harlem near the borderline between black, Hispanic, and white neighborhoods at 110th Street, which of these populations or combination of them is appropriate for calculating the chance that a "random" person would match the DNA found at the scene of the crime? A paradigm case was tried last year in Franklin County, Vermont. DNA from blood stains found at the scene of a lethal assault matched the DNA of an accused man. The prosecution compared the pattern with population samples of various racial groups, and claimed that the chance that a random person other than the accused would have such a pattern was astronomically low.

Franklin County, however, has the highest concentration of Abenaki Indians and Indian/European admixture of any county in the state. The Abenaki and Abenaki/French Canadian population are a chronically poor and underemployed sector in rural Franklin County and across the border in the St. Jacques River region of Canada, where they have been since the Western Abenaki were resettled in the eighteenth century. The victim, like the accused, was half Abenaki, half French-Canadian and was assaulted where she lived, in a trailer park, about one third of whose residents are of Abenaki ancestry. It is a fair presumption that a large fraction of the victim's circle of acquaintance came from the Indian population. No information exists on the frequency of DNA patterns among Abenaki and Iroquois, and on this basis the judge excluded the DNA evidence. But the state could easily argue that a trailer park is open to access from any passerby and that the general population of Vermont is the appropriate base of comparison. Rather than objective science we are left with intuitive arguments about the patterns of people's everyday lives.

The dream of the prosecutor, to be able to say, "Ladies and gentlemen of the jury, the chance that someone other than the defendant could be the criminal is 1 in 3,426,327" has very shaky support. When biologists have called attention to the weaknesses of the method in court or in scientific publications they have been the objects of considerable pressure. One author was called twice by an agent of the Justice Department, in what the scientist describes as intimidating attempts to have him withdraw a paper in press.[3] Another was asked questions about his visa by an FBI agent attorney when he testified, a third was asked by a prosecuting attorney how he would like to spend the night in jail, and a fourth received a fax demand from a federal prosecutor requiring him to produce peer reviews of a journal article he had submitted to the *American Journal of Human Genetics*, fifteen minutes before a fax from the editor of the journal informed the author of the existence of the reviews and their contents. Only one of the authors discussed here, Christopher Wills, discusses the forensic use of DNA, and he has been a prosecution witness himself. He is dismissive of the problems and seems to share with prosecutors the view that the nature of the evidence is less important than the conviction of the guilty.

Both prosecutors and defense forces have produced expert witnesses of considerable prestige to support or question the use of DNA profiles as a forensic tool. If professors from Harvard disagree with professors from Yale (as in this case), what is a judge to do? Under one legal precedent, the so-called "*Frye* rule,"[4] such a disagreement is cause for barring the evidence which "must be sufficiently established to have gained general acceptance in the particular field in which it belongs." But all jurisdictions do not follow *Frye*, and what is "general acceptance," anyway? In response to mounting pressure from the courts and the Department of Justice, the National Research Council was asked to form a Committee on DNA Technology in Forensic Science, to produce a definitive report and recommendations. They have now done so, adding greatly to the general confusion.[5]

Two days before the public release of the report, *The New York Times* carried a front-page article by one of its most experienced

and sophisticated science reporters, announcing that the NRC
Committee had recommended that DNA evidence be barred from
the courts. This was greeted by a roar of protest from the committee, whose chairman, Victor McKusick of Johns Hopkins University, held a press conference the next morning to announce that
the report, in fact, approved of the forensic use of DNA substan-
tially as it was now practiced. The *Times*, acknowledging an
"error," backed off a bit, but not much, quoting various experts
who agreed with the original interpretation. A member of the
committee was quoted as saying he had read the report "fifty
times" but hadn't really intended to make the criticisms as strong
as they actually appeared in the text.

One seems to have hardly any other choice but to read the
report for oneself. As might be expected the report says in effect,
"none of the above," but in substance it gives prosecutors a pretty
tough row to hoe. Nowhere does the report give wholehearted
support to DNA evidence as currently used. The closest it comes
is to state:

> The current laboratory procedure for detecting DNA varia-
> tion . . . is *fundamentally* sound [emphasis added]. . . . It is
> now clear that DNA typing methods are a most powerful
> adjunct to forensic science for personal identification and
> have immense benefit to the public.

and further that

> DNA typing is capable, *in principle*, of an extremely low
> inherent rate of false results [emphasis added].

Unfortunately for the courts looking for assurances, these statements are immediately preceded by the following:

> The committee recognizes that standardization of practices
> in forensic laboratories in general is more problematic than
> in other laboratory settings; stated succinctly, forensic scien-

tists have little or no control over the nature, condition, form, or amount of sample with which they must work.

Not exactly the ringing endorsement suggested by Professor McKusick's press conference. On the other hand, there are no statements calling for the outright barring of DNA evidence. There are, however, numerous recommendations which, taken seriously, will lead any moderately businesslike defense attorney to file an immediate appeal of any case lost on DNA evidence. On the issue of laboratory reliability the report says:

Each forensic-science laboratory engaged in DNA typing must have a formal, detailed quality-assurance and quality-control program to monitor work.

and

Quality-assurance programs in individual laboratories alone are insufficient to ensure high standards. External mechanisms are needed. . . . Courts should require that laboratories providing DNA typing evidence have proper accreditation for each DNA typing method used.

The committee then discusses mechanisms of quality control and accreditation in greater detail. Since no laboratory currently meets those requirements and no accreditation agency now exists, it is hard to see how the committee's report can be read as an endorsement of the current practice of presenting evidence. On the critical issue of population comparisons the committee actually uses legal language sufficient to bar any of the one-in-a-million claims that prosecutors have relied on to dazzle juries:

Because it is impossible or impractical to draw a large enough population to test directly calculated frequencies of any particular profile much below 1 in 1,000, there is not a sufficient body of empirical data on which to base a claim that such frequency calculations are reliable or valid.

"Reliable" and "valid" are terms of art here and Judge Jack Weinstein, who was a member of the committee, certainly knew that. This sentence should be copied in large letters and hung framed on the wall of every public defender in the United States. On balance, *The New York Times* had it right the first time. Whether by ineptitude or design the NRC Committee has produced a document rather more resistant to spin than some may have hoped.

In order to understand the committee's report, one must understand the committee and its sponsoring body. The National Academy of Sciences is a self-perpetuating honorary society of prestigious American scientists, founded during the Civil War by Lincoln to give expert advice on technical matters. During the Great War, Woodrow Wilson added the National Research Council as the operating arm of the Academy, which could not produce from its own ranks of eminent ancients enough technical competence to deal with the growing complexities of the government's scientific problems. Any arm of the state can commission an NRC study and the present one was paid for by the FBI, the NIH Human Genome Center, the National Institute of Justice, the National Science Foundation, and two non-federal sources, the Sloan Foundation and the State Justice Institute.

Membership in study committees almost inevitably includes divergent prejudices and conflicts of interest. The Forensic DNA Committee included people who had testified on both sides of the issue in trials and at least two members had clear financial conflicts of interest. One was forced to resign near the end of the committee's deliberations when the full extent of his conflicts was revealed. A preliminary version of the report, much less tolerant of DNA profile methods, was leaked to the FBI by two members of the committee, and the Bureau made strenuous representations to the committee to get them to soften the offending sections. Because science is supposed to find objective truths that are clear to those with expertise, NRC findings do not usually contain majority and minority reports, and, of course, in the present case a lack of unanimity would be the equivalent of a negative verdict. So we may expect reports to contain contradictory compromises among contending interests, and public pronouncements about a report may be in contradiction to its effective content. *DNA Technology in Forensic Science* in its formation and content is a gold-mine for the serious student of political science and scientific politics.

There is no aspect of our lives, it seems, that is not within the territory claimed by the power of DNA. In 1924, William Bailey published in *The Washington Post* an article about "radithor," a radioactive water of his own preparation, under the headline, "Science to Cure All the Living Dead. What a Famous Savant has to Say about the New Plan to Close Up the Insane Asylums, Wipe Out Illiteracy, and Make over the Morons by his Method of Gland Control."[6] Nothing was more up-to-date in the 1920s than a combination of radioactivity and glands. Famous savants, it seems, still have access to the press in their efforts to sell us, at a considerable profit, the latest concoction.

A STORY IN
TEXTBOOKS

*T*he claim that all of human existence is controlled by our DNA is a popular one. It has the effect of legitimizing the structures of society in which we live, because it does not stop with the assertion that the differences in temperament, ability, and physical and mental health between us are coded in our genes. It also claims that the political structures of society — the competitive, entrepreneurial, hierarchical society in which we live and which differentially rewards different temperaments, different cognitive abilities, and different mental attitudes — is also determined by our DNA, and that it is, therefore, unchangeable. For after all, even if we were biologically different from one another, that in itself would not guarantee that society would have given different power and status to people who are different. That is, to make the ideology of biological determinism complete, we have to have a theory of unchangeable human nature, a human nature that is coded in our genes.

Every political philosophy has to begin with a theory of human nature. Surely, if we cannot say what it is to be truly human, we cannot argue for one or another form of social organization. Social revolutionaries, especially, must have a notion of what it is to be truly human because the call for revolution is the call for the spilling of blood and a wholesale reorganization of the world. One cannot call for a violent overthrow of what is, without claiming that what will be is more in accord with the true nature of human existence. So, even Karl Marx, whose view of society was an historical one, nevertheless believed that there was a true human nature and that human beings realize themselves in their essence by a planned social manipulation of nature for human welfare.

The problem for political philosophers has always been to try to justify their particular view of human nature. Before the seventeenth century, the appeal was made to divine wisdom. God had made people in a certain way. Indeed, they were made in God's image, although a rather blurred one, and moreover, human beings were basically sinful from the time of Adam and Eve's

Fall. But modern secular technological society cannot draw its political claims from divine justification. From the seventeenth century onward, political philosophers have tried to create a picture of human nature based on some sort of appeal to a naturalistic view of the world. Thomas Hobbes in his *Leviathan*, which argued for the necessity of the king, built a picture of human nature from the simplest axioms about the nature of human beings as organisms. To Hobbes, human beings, like other animals, were self-enlarging, self-aggrandizing objects that simply had to grow and occupy the world. But the world was a place of finite resources, and so it necessarily would happen that human beings would come into conflict over those resources as they expanded and the result would be what he called "the war of all against all." The conclusion for Hobbes was that one needed a king to prevent this war from destroying everything.

The claims that organisms, especially human beings, grow without bound and that the world in which they grow is finite and limited are the two basic claims that have given rise to the modern biological theory of human nature. They resurfaced in the Reverend Malthus's treatise on population, in his famous law that organisms grow geometrically in numbers while the resources for their subsistence grow only arithmetically, and so again a struggle for existence must occur. As we all know, Darwin took over this notion of nature to build his theory of natural selection. Since all organisms are engaged in a struggle for existence, those that are better suited by their shape and form, by their physiology, and by their behavior to leave more offspring in that struggle will do so, and the consequence will be that their kind will take over the earth. The Darwinian view is that whatever human nature may be, it, like everything else about humans, who are after all living organisms, has evolved by natural selection. Therefore, what we truly are is the result of two billions years of evolution from the earliest rudimentary organisms to us.

As evolutionary theory has developed over the last one hundred years and become technologically and scientifically sophisticated, as vague notions of inheritance have become converted into a very precise theory of the structure and function of DNA, so the evolutionary view of human nature has developed a modern, scientific-sounding apparatus that makes it seem every bit as unchallengeable as the theories of divine providence seemed in an earlier age. What has happened in effect is that Thomas Hobbes's war of all against all has been converted into a struggle between DNA molecules for supremacy and dominance over the structures of human life.

The most modern form of naturalistic human nature ideology is called sociobiology. It emerged onto the public scene about 15 years ago and has since become the ruling justifying theory for the permanence of society as we know it. It is an evolutionary and a genetic theory that uses the entire theoretical apparatus of modern evolutionary biology, including a great deal of abstruse mathematics, which is then translated for the inexpert reader in coffee-table books with beguiling pictures and in magazine articles and newspaper accounts. Sociobiology is the latest and most mystified attempt to convince people that human life is pretty much what is has to be and perhaps even ought to be.

The sociobiological theory of human nature is built in three steps. The first is a description of what human nature is like. One looks around at human beings and tries to build a fairly complete description of the features that are said to be common to all human beings in all societies in all places in all times.

The second step is to claim that those characteristics that appear to be universal in humans are, in fact, coded in our genes, that is, in our DNA. There are genes for religiosity, genes for entrepreneurship, genes for whatever characteristics are said to be built into the human psyche and human social organization.

These two claims — that there is a universal human nature and that it is coded in the genes and is unchangeable — would be sufficient as a biological theory of human nature in a purely descriptive sense. That is what we are; take it or leave it. But sociobiological theory, being built on evolutionary theory, goes one step further, as it must to fulfill its program. It must explain, and in some sense justify, how we come to have these particular genes rather than some other genes that might have given us quite a different human nature.

The theory thus goes on to the third step, the claim that natural selection, through the differential survival and reproduction of different kinds of organisms, has led inevitably to the particular genetic characteristics of individual human beings, characteristics that are responsible for the form of society. This claim strengthens the argument of legitimacy because it goes beyond mere description to assert that the human nature described is inevitable, given the universal law of the struggle for existence and the survival of the fittest. In this sense, the sociobiological theory of human nature puts on a mantle of universality and of utter fixity. After all, if 3 billion years of evolution have made us what we are, do we really think that a hundred days of revolution will change us?

Sociobiologists take the first step, the claimed correct description of what is universal in all human beings, more or less as every human nature theorist has done it, by looking around to see what people in their society are like and to some extent by telling their own life stories. Having looked inward at themselves and outward at modern capitalist society for a description of human nature, they then extend it a bit further by looking into the anthropological record in order to assure us that those very same elements that they find in twentieth-century North America and Britain are also, in one form or another, displayed by the Stone Age people of New Guinea. For some reason, they do not look much at the historical record of European society, of which they seem to be quite ignorant, but perhaps they feel that if New Guinea highlanders and Scottish highlanders show the same characteristics today, then there cannot have been much change in 1,500 years of recorded history.

And what are these human universals that sociobiologists find? One can hardly do better than look at the most influential, and in some sense, founding document of sociobiological theory, E.O. Wilson's *Sociobiology: The New Synthesis*.[1]

Professor Wilson tells us, for example, that human beings are indoctrinable. He says, "Human beings are absurdly easy to indoctrinate. They seek it."[2] They are characterized by blind faith: "Man would rather believe than know."[3] That statement is, we must note, found in what is called a scientific work, used as a textbook in courses all over the world, filled with the mathematics of modern population biology, crammed with observations and facts about the behavior of all kinds of animals, based on what *Time* magazine has called the "iron laws of nature." But surely, "man would rather believe than know" is more in the line of barroom wisdom, the sort of remark one makes to one's friend at the local after work following a particularly frustrating attempt to persuade the person in the next office that he ought to do things in a different way. Among other aspects of human nature are said to be a universal spite and family chauvinism. We are told that "human beings are keenly aware of their own blood lines and the intelligence to plot intrigue."[4] Xenophobia, the fear of strangers, is part of our universal equipment. "Part of man's problem is that his intergroup responses are still crude and primitive and inadequate for the extending territorial relationship that civilization has thrust upon him."[5] One of the results of this, we are told, is that "the most distinctive human qualities have emerged during the phase of social evolution that occurred through intertribal warfare and through genocide."[6] And then there is the relationship between the sexes. Male dominance and superiority is part of human nature. Wilson writes that "among general social traits in human beings are aggressive dominance systems with males dominant over females."[7] The list is not complete. Nor is this simply the idiosyncratic view of one influential sociobiologist. The claims that human warfare, sexual dominance, love of private property, and hate of strangers are human universals are found over and over in the writings of sociobiologists, whether they be biologists, economists, psychologists, or political scientists.

But to make such claims, one must be quite blind even to the history of European society. Take, for example, the claim of a universal xenophobia. In fact, the attitudes of people toward foreign cultures and other countries have varied tremendously from social class to social class and time to time. Could the aristocracy of Russia in the nineteenth century, which thought all things Slavic to be inferior, which spoke French by preference, which looked to Germany for its military and technological resources, be described as xenophobic? Educated and upper classes in particular have often looked to other cultures for the highest and the best. English-speaking scientists on occasion are interviewed by Italian radio and television, and the answers given to the producer's questions are translated into Italian, which the listeners hear in a voice-over after a few moments of the scientist's English. When the producers are asked why they do not get an Italian scientist to do the program, they say that Italians simply do not believe any claims about science that are made in Italian and that they have to hear it in English if they are to believe it is true.

Nothing better reveals the narrow ahistorical claims of sociobiological description than the standard discussion of the economy of scarcity and unequal distribution. So, Professor Wilson writes that "the members of human societies sometimes cooperate closely in insectan fashion, but more frequently they compete for the limited resources allocated to their role sector. The best and the most entrepreneurial of the role actors usually gain a disproportionate share of the rewards while the least successful are displaced to other less desirable positions."[8] But this description completely ignores the immense amount of sharing of resources that occurs among a whole variety of modern hunting and gathering societies like Eskimos, and it completely distorts the history even of Europe. The concept of entrepreneurship does not work for, say, the thirteenth century in the Île-de-France, an agrarian feudal society in which land could not be bought and sold, in which labor could not be hired and fired, and in which the so-called market mechanism was a rudimentary form of exchange of a few goods. Of course, sociobiologists recognize that there are exceptions to these generalizations, but their claim is that those exceptions are temporary and unnatural, and that they will not persist in the absence of constant force and threat. So, societies may indeed, like blue-clad regimented Chinese, cooperate in so-called insectan fashion. But this can be managed only by constant supervision and force. The moment one relaxes one's vigil, people will revert to their natural ways. It is rather as if we could make a law saying that everyone would have to walk on their knees, which would be physically possible but terribly painful. The moment we relaxed our vigil, everyone would stand upright again.

At the surface of this theory of human nature is the obvious ideological commitment to modern entrepreneurial competitive hierarchical society. Yet underneath is a deeper ideology, and that is the priority of the individual over the collective. Despite the name *socio*biology, we are dealing with a theory not of social causation but of individual causation. The characteristics of society are seen as caused by the individual properties that its members have, and those properties, as we shall see, are said to derive from the members' genes. If human societies engage in war, that is because each individual in the society is aggressive. If men as a group dominate women or whites Blacks, it is because each man as an individual is desirous of dominating each woman and each white person has feelings of personal hostility set off by the sight of Black skin. The structures of society simply reflect these individual predispositions. Society is nothing but the collection of individuals in it, just as culture is seen as nothing but the collection of disarticulated bits and pieces, individual preferences and habits.

Such a view completely confuses, partly by linguistic confusion, very different phenomena. It is obviously not the case that Britain and Germany made war on each other in 1914 because individual Britons and individual Germans felt aggressive. If that were the case, we would not need conscription. Englishmen, Canadians, and Americans killed Germans and vice versa because the state put them in a position that made it inevitable they did so. A refusal to be conscripted meant a jail term and the refusal to obey orders in the field meant death. Great machines of propaganda, martial music, and stories of atrocities are manufactured by the state to convince its citizens that their lives and the chastity of their daughters are at risk in the face of the threat of barbarians. The confusion between individual aggression and national aggression is a confusion between the rush of hormones that may be felt if someone is slapped in the face and a national political agenda to control natural resources, lines of commerce, prices of agricultural goods, and the availability of labor forces that are the origins of warfare. It is important to realize that one does not have to have a particular view of the content of human nature to make this error of individuals causing society. Prince Kropotkin, a famous anarchist, also claimed that there was a universal human nature but one that would create cooperativeness and would be anti-hierarchical if only it were allowed free play.[9] But his theory was no less a theory of the dominance of the individual as the source of the social.

Having described a universal set of human social institutions that are said to be the consequence of individual natures, sociobiological theory then goes on to claim that those individual properties are coded in our genes. There are said to be genes for entrepreneurship, for male dominance, for aggressivity, so conflict between the sexes and parents and offspring is said to be genetically programmed. What is the evidence that these claimed human universals are in fact in the genes? Often, it is simply asserted that because they are universal they must be genetic. A classic example is the discussion of sexual dominance. Professor Wilson has written in *The New York Times*, "In hunter-gatherer societies, men hunt and women stay at home. This strong bias persists in most agricultural and industrial societies [apparently, he has not yet caught up with women in the workforce], and on that ground alone appears to have a genetic origin."[10] This argument confuses the observation with its explanation. If the circularity is not obvious, we might consider the claim that since 99 percent of Finns are Lutherans, they must have a gene for it.

A second evidence offered for the genetic determination of human universal traits is the claim that other animals show the same traits and, therefore, we must have a genetic continuity with them. Ants are described as making "slaves" and having "queens." But the slavery of ants knows nothing of the auction block, of the buying and selling, of the essentially commodity nature of the slave relations of human society. Indeed, ant slaves are almost always of other species, and ant slavery has a great deal more in common with the domestication of animals. Nor do ants have "queens". The force-fed egg factory encased in a special chamber in the middle of an ant colony that is called a queen has no resemblance to the life of either Elizabeth I or Elizabeth II or of their different political roles in society. Nor are the words "slave" and "queen" simply convenient labels. Ant "slavery" and ant "royalty" are claimed to have important causal continuity with their human counterparts. They are said to be products of the same forces of natural selection.

This confusion between qualities of animals and qualities of human society is an example of the problem of *homology* and *analogy*. By homologous traits, biologists mean those properties of organisms that are shared by different species because they have a common biological origin and some common biological genetic ancestry, and they derive from common features of anatomy and development. Even though they look very different and are used for very different purposes, the bones of a human arm and of a bat's wing are homologous because they are anatomically derived from the same structures and influenced by the same genes. On the other hand, a bat's wing and an insect's wing are only analogous. That is, they look superficially alike and they seem to serve the same function, but they have no origin in common at the genetic or morphological level. But analogy is in the eye of the observer. How do we decide that slavery in ants and ant queens are like human slavery and like human royal families? How do we decide that the coyness we see in people is the same as the behavior in animals called coyness? What happens is that human categories are laid on animals by analogy, partly as a matter of convenience of language, and then these traits are "discovered" in animals and laid back on humans as if they had a common origin. There is in fact not a shred of evidence that the anatomical, physiological, and genetic basis of what is called aggression in rats has anything in common with the German invasion of Poland in 1939.

The third kind of evidence that is presented for a genetic basis of human social behavior is the report of heritability of human traits. Such characteristics as introversion and extroversion, personal tempo, psychomotor and sports activities, eroticism, dominance, depression, and even conservatism and liberalism are said to be heritable. But the evidence for the heritability of these traits is totally absent. We must remember that genetics is a study of similarity and difference between relatives. We judge things to be heritable if close relatives are more alike than distant relatives or unrelated persons. But the problem in human genetics in particular is that similarity between relatives arises not only for biological reasons but for cultural reasons as well, since members of the same family share the same environment. This has always been the problem of human genetics whether we are talking about traits of personality or anatomy. Most reports of the heritability of personality traits are simple observations that parents and children resemble each other in some respect. The highest similarity between parents and offspring for social traits in North America is for political party and religious sect, yet no serious person believes genes determine these attributes. The observation of similarity of parents and offspring is not evidence of their biological similarity. There is a confusion between the observation and the possible causes. The fact is, not a single study of personality traits in human populations successfully disentangles similarity because of shared family experience and similarity because of genes. So, in fact, we know nothing about the heritability of human temperamental and intellectual traits that are supposed to be the basis for social organization.

There is a deeper problem. To carry out a heritability study, even a correct one, we require differences between individuals. If everyone is identical in some respect, that is, if everybody has exactly the same genes for some characteristic, then there is no way to investigate its heritability, because genetic investigations require contrasts between individuals. Sociobiological theory claims that all human beings share genes for aggression, for xenophobia, for male dominance, and so on. But if we all share these genes, if evolution has made us all alike in this human nature, then in principle there would be no way to investigate the heritability of the traits. On the other hand, if there is genetic variation among human beings in these respects, then on what basis do we declare that one or another manifestation is universal human nature? If it is genetically determined human nature that we are aggressive and like to go to war, then we must suppose that A.J. Mustie, the famous pacifist, lacked this gene and was, therefore, in some sense less than human. If, on the other hand, he possessed the gene but was a pacifist, the genes seem somewhat less than all-powerful in determining behavior. Why are we not all like A.J. Mustie? There are deep contradictions in simultaneously asserting that we are all genetically alike in certain respects, that our genes are all-powerful in determining our behavior, and at the same time observing that people differ.

Finally, there is an extraordinary biological naiveté and ignorance of the principles of developmental biology involved in assertions that genes make us behave in particular ways in particular circumstances. DNA functions in several ways in influencing the development of organisms.

First, the exact sequence of the amino acids in our proteins is coded in our genes, but no one would suggest that the amino acid sequence for a particular protein in itself can make us liberal or conservative. Second, genes influence when in the course of development and in which part of the body particular proteins are to be produced, and this in turn influences cell division and cell growth. So it might be claimed that there is a fixed pattern of neurons in our central nervous system, influenced by the turning on and turning off of genes during development, that makes

us warlike or pacifist. However, this would require a theory of
the development of the central nervous system that makes no
allowances for developmental accidents and little or no role for
the creation of mental structures by experience. Yet even the rudimentary social organization of ants, with their structure of work
and interindividual relations that is so simple compared with
ours, is very flexible with respect to information for the external
world. Ant colonies change their collective social behavior over
time and depending on how long the colony has occupied a given
territory. It takes an enormous set of assumptions to suppose that
the human central nervous system, with thousands of times more
nervous connections than in an ant, has completely stereotyped
and fixed genetic responses to circumstance. The incredible
variety of human social circumstances would require an amount
of DNA that we simply do not possess. There is enough human
DNA to make about 250,000 genes. But that would be insufficient to determine the incredible complexity of human social
organization if it were coded in detail by specific neuronal connections. Once we admit that only the most general outlines of
social behavior could be genetically coded, then we must allow
immense flexibility depending on particular circumstances.

   The final step in the sociobiological argument is to say that
the genes we possess for universal human nature have been
established in us through evolution by natural selection. That is,
once upon a time human beings varied genetically in the degree
to which they were aggressive, xenophobic, indoctrinable, male
dominant, and so on, but those individuals who were most
aggressive or most male dominant left more offspring, so the
genes that were eventually left in us as a species were the ones
that now determine those traits. The argument of natural selection seems a fairly simple and straightforward one for some kinds
of traits. For example, it is argued, the more aggressive of our
ancestors would leave more offspring because they would swoop
down on the less aggressive and eliminate them. The more entrepreneurial would have appropriated more resources in short
supply and starved out the wimps. In each of these cases, it is easy

to make up a plausible story that would explain the superior reproductive abilities of one type over another.

There are, however, some traits that are said to be universal and that do not lend themselves so easily to this story of individual reproductive advantage. An example, and one that is discussed a great deal by sociobiologists, is altruistic behavior. Why should we be cooperative under some circumstances, and why should we sometimes give up what appear to be immediate advantages for the benefit of others? To explain altruism, sociobiologists advance the theory of kin selection. Natural selection for a trait does not require that individuals possessing it leave more offspring but only that the genes coding the trait be represented in larger numbers in future generations.

There are two ways to increase the representation of one's genes in future generations. One is to leave more offspring. The other is to arrange that even if one does not leave more offspring, one's relatives do so, since close relatives share genes. So, a person could sacrifice his reproduction completely, provided his brothers and sisters left many more children. Thus, his kind of genes would increase indirectly through his relatives and, in this indirect way, he would leave more offspring. An example of this phenomenon is the occurrence of "helpers at the nest" in birds, in which it is said that nonreproductive birds help out their close relatives, who are then able to raise more than the ordinary number of offspring and in the end more family genes are left. To make kin selection work, a sufficient number of excess offspring must be left by relatives. For example, if an individual gives up its own reproduction, its brothers and sisters must have twice as many offspring as ordinarily, but one can at least tell a story that might make this plausible.

We are then left with those traits that do not even benefit relatives differentially, for example, a general altruism toward all members of the species. Why are we good to strangers? For this phenomenon, the sociobiologist provides the theory of "reciprocal altruism". The argument is that even if we are unrelated, if I do you a favor that costs me something, you will remember that favor and reciprocate in the future, and by this indirect path I will succeed in advancing my own reproduction. An example often given is that of the drowning person. You see someone drowning and jump in to save that person even at the risk of your own life. In the future, when you are drowning, the person whose life you have saved will remember, and save you in gratitude. By this indirect path you will increase your own probability of survival and reproduction over the long run. The problem with this story is, of course, that the last person in the world you want to depend on to save you when you are drowning is someone whom you had to save in the past, since he or she is not likely to be a strong swimmer.

The real difficulty with the process of explanation that allows direct advantage, or kin selection, or reciprocal altruism when one or the other is useful in the explanation, is that a story can be invented that will explain the natural selective advantage of any trait imaginable. When we combine individual selective advantage with the possibility of kin selection and reciprocal altruism, it is hard to imagine any human trait for which a plausible scenario for its selective advantage could not be invented. The real problem is to find out whether any of these stories is *true*. One must distinguish between plausible stories, things that *might* be true, and true stories, things that actually have happened. How do we know that human altruism arose because of kin selection or reciprocal altruistic selection? At the very minimum, we might ask whether there is any evidence that such selective processes are going on at the present, but in fact no one has ever measured in any human population the actual reproductive advantage or disadvantage of any human behavior. All of the sociobiological explanations of the evolution of human behavior are like Rudyard Kipling's *Just So* stories of how the camel got his hump and how the elephant got his trunk. They are just stories. Science has been turned into a game.

The entire process of sociobiological reasoning is illustrated by two cases, one meant only fancifully by sociobiologists as a teaching exercise, and one they take seriously. The fanciful one concerns the problem of why children hate spinach and adults like it. It is contained in a high-school textbook written by sociobiologists in order to train children in adaptive thinking.[11] The first step is the description of a human universal. All children hate spinach. To see the universal truth of this assertion, it is only necessary to look around and ask our friends. Moreover, when we look around we find that adults eat spinach. How has this happened? We imagine that there is a gene that causes children to hate spinach but allows adults to like it. Note that there is no evidence for such an unlikely gene. It is simply postulated. Spinach contains a substance, oxalic acid, that interferes with the absorption of calcium, which young children need for their growing bones. So any child in the past who had the wrong gene and ate spinach would have had defective growth, suffered from rickets, and might not have left many offspring. (Although it is far from clear that crooked legs interfere with reproduction and longevity.) On the other hand, adults' bones have stopped growing, calcium is not so important to them, and they are free to take advantage of the nutrients in spinach, so there is no selection against their liking it. As a consequence, it is part of genetically determined human nature that children hate spinach but adults like it. We have a completely articulated story of a claimed universal fact of human nature. We should not let the silliness of this case distract from its essential features. It is meant to teach students all the elements of a naturalistic argument about human nature. It makes a generalized observation by looking around. It postulates genes without any evidence, and then it tells a plausible or perhaps not so plausible story.

Let us see how this is applied to a serious case and one widely discussed by sociobiologists: the existence in human societies of homosexuality. Homosexuality is claimed to be a biological problem because, after all, since homosexuals leave no offspring, the genes for homosexuality should have long ago disappeared. Why have they not? First, the sociobiologist makes the assumption that homosexuals leave fewer offspring. This implies a description of human sexual behavior in which the world is divided between heterosexuals and homosexuals, one class that leaves offspring and the other that does not. This description, however, does not correspond to our knowledge of human sexuality. In fact, the world is not divided between two classes. On the contrary, there is continuum of sexuality from persons who have never engaged in anything but heterosexual behavior, through those who have a somewhat wider range of experiences, through those who are regularly bisexual to those who are totally homosexual. According to a number of surveys, about half of all males in North America have had at least one homosexual contact. Moreover, this range of behaviors has varied historically by social class. There was widespread bisexuality among the upper classes in Classical Rome and Greece and, indeed, what were the usual homosexual practices in these societies were different from present practices. Moreover, there is, curiously enough, not an iota of evidence about the relative reproductive rates of people with different sexual histories. Obviously those who engage *exclusively* in homosexual behavior have, until the recent advent of artificial insemination, left no offspring. But nothing is known about the reproductive rates of those who are totally heterosexual as opposed to those who have a broad range of sexual experience. So, for example, we do not know whether a person who is heterosexual in 40 percent of his or her encounters and homosexual in 60 percent has fewer or more children than a person who is totally heterosexual. Indeed, we could make an argument that bisexuality is a manifestation of greater general libido, and it might turn out that bisexual people leave more offspring. We simply do not know the answer.

Second, there is absolutely no evidence that there are any genetic differences between individuals of different sexual preferences. If it were true that homosexuals left no offspring, there would be a certain problem in studying the heritability of homosexual behavior. In fact, there are no studies of the heritability of sexual preference, so the claims for genetic predisposition to different forms of sexuality are pure fancy.

Finally, there is the problem of an evolutionary story. The one told by sociobiologists is that of helpers at the nest. Homosexuals in the past, it is argued, did not themselves leave any offspring but helped their heterosexual brothers and sisters to raise more children by sharing resources with them, and this compensation was sufficient to keep the genes for homosexuality in the population. It must be remembered that the nonreproductive homosexuals must help their brothers and sisters so well that those relatives have twice as many offspring as usual if kin selection is to work. But there is no evidence for helpers at the nest in human societies. If our remote prehistoric ancestors were anything like modern hunting and gathering people, a general sharing of resources would be a common phenomenon not only within the family but within entire villages so that the "nest" included nonrelatives. But we know nothing about the relative number of offspring left at present by people who have homosexual brothers and sisters, for no one has ever measured family size in relation to this question.

Thus, the entire discussion of the evolutionary basis of human sexual preference is a made-up story, from beginning to end. Yet it is a story that appears in textbooks, in courses in high schools and universities, and in popular books and journals. It bears the legitimacy given to it by famous professors and by national and international media. It has the authority of science. In an important sense, it *is* science because science consists not simply of a collection of true facts about the world, but is the body of assertions and theories about the world made by people who are called scientists. It consists, in large part, of what scientists say about the world whatever the true state of the world might be.

Science is more than an institution devoted to the manipulation of the physical world. It also has a function in the formation of consciousness about the political and social world. Science in that sense is part of the general process of education, and the assertions of scientists are the basis for a great deal of the enterprise of forming consciousness. Education in general, and scientific education in particular, is meant not only to make us competent to manipulate the world but also to form our social attitudes. No one saw this more clearly, and was more honest about it, than one of the most conservative political figures in American history, Daniel Webster, who wrote that "education is a wise and liberal form of police by which property and life and the peace of society are secured."

SCIENCE AS
SOCIAL ACTION

*T*he previous pages have all been concerned with a particular ideological bias of modern biology. That bias is that everything that we are, our sickness and health, our poverty and wealth, and the very structure of the society we live in are ultimately encoded in our DNA. We are, in Richard Dawkins's metaphor, lumbering robots created by our DNA, body and mind. But the view that we are totally at the mercy of internal forces present within ourselves from birth is part of a deep ideological commitment that goes under the name of *reductionism*. By reductionism we mean the belief that the world is broken up into tiny bits and pieces, each of which has its own properties and which combine together to make larger things. The individual makes society, for example, and society is nothing but the manifestation of the properties of individual human beings. Individual internal properties are the causes and the properties of the social whole are the effects of those causes. This individualistic view of the biological world is simply a reflection of the ideologies of the bourgeois revolutions of the eighteenth century that placed the individual at the center of everything.

Such a view about causes and effects and the autonomy of individual bits and pieces not only results in a belief that internal forces beyond our control govern what we are as individuals. It also posits an external world with its own bits and pieces, its own laws, which we as individuals confront but do not influence. Just as the genes are totally inside of us, so the environment is totally outside of us, and we as actors are at the mercy of both these internal and external worlds. This gives rise to the false dichotomy of nature and nurture. Against those who say that our ability to solve problems, our intelligence, is determined by our genes, there exists a contrary party that claims that our intelligence is determined by our environment. And so the struggle goes on between those who believe in the primacy of nature and those who believe in the primacy of nurture.

The separation between nature and nurture, between the organism and the environment, goes back to Charles Darwin, who finally brought biology into the modern mechanistic world view. Before Darwin, it was the general view that what was outside and what was inside were part of the same whole system and one could influence the other. The most famous theory of evolution before Darwin was that of Jean Baptiste Lamarck, who believed in the inheritance of acquired characteristics. Changes occurred in the environment that caused changes in the body or behavior of organisms, and it was believed that the changes induced by the environment would enter into the hereditary structure of the organisms and would be passed on to the next generation. In this view, nothing separates what is outside from what is inside because external alterations would enter into the organism and be perpetuated in future generations.

Darwin completely rejected this world view and replaced it with one in which organisms and environment were totally separated. The external world had its own laws, its own mechanisms of operation. Organisms confronted these and experienced them and either successfully adapted to them or failed. The rule of life, according to Darwin, is "adapt or die." Those organisms whose properties enabled them to cope with the problems set by the external world would survive and leave offspring, and the others would fail to do so. The species would change, not because the environment directly caused physical and body changes in organisms, but because those organisms smart enough to be able to handle the problems thrown at them by nature would leave more offspring, who would resemble them. The deep point of Darwinism was the separation between the forces of the environment that create the problems and the internal forces of the organism that throw up solutions to problems more or less at random, the correct solutions being preserved. The external and internal forces of the world behave independently. The only connection between them is a passive one. The organisms who happen to be lucky enough to find a match between what is going on inside themselves and what was going on outside themselves survive.

Darwin's view was essential to our successful unraveling of evolution. Lamarck was simply wrong about the way the environment influences heredity, and Darwin's alienation of the organism from the environment was an essential first step in a correct description of the way the forces of nature act on each other. The problem is that it was only a *first* step, and we have become frozen there. Modern biology has become completely committed to the view that organisms are nothing but the battle grounds between the outside forces and the inside forces. Organisms are the passive consequences of external and internal activities beyond their control. This view has important political reverberations. It implies that the world is outside our control, that we must take it as we find it and do the best we can to make our way through the mine field of life using whatever equipment our genes have provided to us to get to the other side in one piece.

What is so extraordinary about the view of an external environment set for us by nature, and essentially unchangeable except in the sense that we might ruin it and destroy the delicate balance that nature has created in our absence, is that it is completely in contradiction to what we know about organisms and environment. When we free ourselves of the ideological bias of atomism and reductionism and look squarely at the actual relations between organisms and the world around them, we find a much richer set of relations, relations that have very different consequences for social and political action than are usually supposed, for example, by the environmental movement.

First, there is no "environment" in some independent and abstract sense. Just as there is no organism without an environment, there is no environment without an organism. Organisms do not experience environments. They create them. They construct their own environments out of the bits and pieces of the physical and biological world and they do so by their own activities. Are the stones and the grass in my garden part of the environment of a bird? The grass is certainly part of the environment of a phoebe that gathers dry grass to make a nest. But the stone around which the grass is growing means nothing to the phoebe. On the other hand, the stone is part of the environment of a thrush that may come along with a garden snail and break the shell of the snail against the stone. Neither the grass nor the stone are part of the environment of a woodpecker that is living in a hole in a tree. That is, bits and pieces of the world outside of these organisms are made relevant to them by their own life activities. If grass is used to make a nest, then grass is part of the environment. If stones are used to break snails on, then stones are part of the environment.

There is an infinity of ways in which parts of the world can be assembled to make an environment, and we can know what the environment of an organism is only by consulting the organism. Not only do we consult the organism, but when we describe the environment we describe it in terms of the organism's behavior and life activities. If you are in any doubt of this, you might try asking a professional ecologist to describe the environment of some bird. He or she will say something like the following. "Well, the bird builds its nest three feet off the ground in hardwoods. It eats insects part of the year but then may switch to seeds and nuts when insects are no longer available. It flies south in the winter and comes back north in the summer, and when it is foraging for its food it tends to stay in the higher branches and at their outer tips," and so on. Every word uttered by the ecologist in describing the environment of a bird will be a description of the life activities of the bird. That process of description reflects the fact that the ecologist has learned what the environment of the bird is by watching birds.

A practical demonstration of the difficulty of describing an environment without having seen an organism that determined and defined it is the case of the Mars Lander. When the United States decided to send a landing module to Mars, biologists wanted to know if there was any life there. So the problem was to design a machine to detect life on Mars. There were several interesting suggestions. One was to send a kind of microscope with a long sticky tongue that would unroll on the planet's surface and then roll back up and put whatever dust it found under the microscope. If there was anything that looked like a living organism, we would see it in the images sent back to Earth. One might call this the morphological definition of life. If it looks right and it wiggles, then it is alive.

What appears to be a more sophisticated approach was taken. Instead of asking whether things on Mars look alive, it was decided to ask whether they have the metabolism of living things. So the Mars Lander contained what was essentially a long hose attached to a vacuum cleaner inside of which was a container of radioactive growth medium. When the Lander got to Mars, it would suck up some dust into the medium and if there were any living organisms in the dust, they would break down the medium as bacteria do on Earth, radioactive carbon dioxide would be produced, and a detector in the machine would signal the presence of this gas. And that is exactly what happened. When the Mars Lander sucked up the dust, radioactive carbon dioxide was produced in a pattern that had everyone convinced that there was life on Mars fermenting the medium. But then suddenly the process shut down and there was no further fermentation. This was not what living organisms were supposed to do, and the consequence was scientific confusion. After a debate among those concerned with the experiment, it was decided that there was no life on Mars. Instead, it was postulated that there was a kind of chemical reaction on finely divided clay particles catalyzed by the particles, which were not ordinarily seen on Earth. Later, this reaction was successfully mimicked in the laboratory, so everybody has now agreed that they decided correctly and that there is no life on Mars.

The problem with this experiment arises precisely from the fact that organisms define their own environment. How can we know whether there is life on Mars? We present Martian life with an environment and see whether it can live in it. But how can we know what the environment of Martian life is unless we have seen Martian organisms? All that the experiment of the Mars Lander showed was that there is no Earth-like bacterial life on Mars. We may know the temperature, the gas content of the atmosphere, the humidity, and something about the soil on

Mars, but we do not know what a Martian environment is like because the environment does not consist of temperature, gas, moisture, and soil. It consists of an organized set of relationships among bits and pieces of the world, which organization has been created by living Martian organisms themselves.

We must replace the adaptationist view of life with a constructionist one. It is not that organisms find environments and either adapt themselves to the environments or die. They actually *construct* their environment out of bits and pieces. In this sense, the environment of organisms is coded in their DNA and we find ourselves in a kind of reverse Lamarckian position. Whereas Lamarck supposed that changes in the external world would cause changes in the internal structures, we see that the reverse is true. An organism's genes, to the extent that they influence what that organism does in its behavior, physiology, and morphology, are at the same time helping to construct an environment. So, if genes change in evolution, the environment of the organism will change too.

Consider the immediate environment of a human being. If one takes motion pictures of a person, using schlieren optics that detect differences in the refractive index of the air, one can see that a layer of warm, moist air completely surrounds each one of us and is slowly rising from our legs and bodies and going off the top of our heads. In fact, every living organism including trees has this boundary layer of warm air that is created by the organism's metabolism. The result is that we are encapsulated in a little atmosphere created by our own metabolic activities. One consequence is what is called the wind-chill factor. The reason that it gets much colder when the wind blows across us is because the wind is blowing away the boundary layer and our skins are then exposed to a different set of temperatures and humidities. Consider a mosquito feeding on the surface of the human body. That mosquito is completely immersed in the boundary layer that we have constructed. It is living in a warm, moist world. Yet one of the most common evolutionary changes for all organisms is a change in size, and over and over again organisms have evolved to be larger. If the mosquito species begins to evolve to a larger size, it may in fact find itself with its back in the "stratosphere" and only up to its knees in the warm, moist boundary layer while it is feeding. The consequence will be that the mosquito's evolution has put it into an entirely different world. Moreover, as human beings early in their evolution lost hair and the distribution of sweat glands over their bodies changed, the thickness of the boundary layer changed and so changed the micro-world that they carry with them, making it rather less hospitable for fleas, mosquitoes, and other parasites that live on hairy animals. The first rule of the real relation between organisms and environment is that environments do not exist in the absence of organisms but are constructed by them out of bits and pieces of the external world.

The second rule is that the environment of organisms is constantly being remade during the life of those living beings. When plants send down roots, they change the physical nature of the soil, breaking it up and aerating it. They exude organic molecules, humic acids, that change the soil's chemical nature as well. They make it possible for various beneficial fungi to live together with them and penetrate their root systems. They change the height of the water table by removing water. They alter the humidity in their immediate neighborhood, and the upper leaves of a plant change the amount of light that is available to the lower leaves. When the Canadian Department of Agriculture takes weather records for agricultural purposes, they do not set up a weather station in an open field or on the roof of a building. They take measurements of temperature and humidity at various levels above the ground in a field of growing plants because the plants are constantly changing the physical conditions that are relevant to agriculture. Moles burrow in the soil. Earthworms through their castings completely change the local topology. Beavers have had at least as important an effect on the landscape in North America as humans did until the beginning of the last century. Every breath you take removes oxygen and adds carbon dioxide to the world.

Mort Sahl once said, "Remember, no matter how cruel and nasty and evil you may be, every time you take a breath you make a flower happy."

Every living organism is in a constant process of changing the world in which it lives by taking up materials and putting out others. Every act of consumption is also an act of production. And every act of production is an act of consumption. When we consume food, we produce not only gases but solid waste products that are in turn the materials for consumption of some other organism.

A consequence of the universality of environmental change induced by the life activity of organisms is that every organism is both producing and destroying the conditions of its existence. There is a great deal of talk about how we as human beings are destroying the environment. But we are not unique in the fact that our life processes are recreating the world in a way that is in part hostile to the continuation of our own lives. Every bacterium uses up food material and excretes waste products that are toxic to it. Organisms ruin the world not only for their own lives but for their children as well.

The entire vegetational landscape of New England is a consequence of that process. The primeval forest in New England consisted of a mixture of hardwoods, pines and hemlocks. As agriculture spread at the end of the eighteenth and through the nineteenth century, all these forests were cut down and replaced by farms. Then, just before and after the Civil War, there were wholesale migrations out of the rocky soils of New England, where one could barely plant a crop, to the deep and productive soils of the Middle West. As a result, farms were abandoned and plants started to infiltrate these old fields. The first thing that came in was a variety of weeds and herbs. These were replaced later by white pines. White pines can form an almost pure stand in an old field and many such pure white pine stands could be seen in New England earlier in this century. However, they do not last. The pines make a dense shade that is inhospitable to the growth of their own seedlings, and so they cannot replace each other. As the pines die or if, as in New England, they are cut wholesale, what comes in are hardwoods, whose seedlings have been waiting around for a little opening. The white pines disappear forever with the exception of an occasional old tree, and a composition similar to the prehistoric virgin forest appears. This old-field white pine to hardwood succession is a consequence of the conditions of light and soil being changed by the pine trees in such a way that their own offspring cannot succeed them. The generation gap is not simply a human phenomenon.

So, we must put away the notion that out there there is a constant and fixed world that human beings alone are disturbing and destroying. We are certainly changing it, as all organisms do, and we certainly have a power that other organisms do not have, both to change the world extremely rapidly and, by willful activity, to change the world in various ways that we may think beneficial. Nevertheless, we cannot live without changing the environment. That is the second law of the relationship between organism and environment.

Third, organisms determine the statistical nature of the environment at least as far as it has an influence on themselves. Organisms are capable of averaging over time and buffering out the fluctuations in physical factors. An important example is the way animals and plants store sunlight. Even though the conditions for growth and good nutrition do not exist all year around in a temperate zone, it is not only farmers who make hay while the sun shines. Potatoes are the storage organs of potato plants and acorns form the storage for oak trees. Other organisms, in turn, use these storage devices for their own storage. Squirrels store away acorns for use in the winter and human beings store away potatoes. As human beings, we have even a further level of averaging: money. Money is the way in which, through futures contracts, fluctuations in the availability of natural products are ironed out for the market, and savings banks are where we put money for a rainy day. So organisms do not, in fact, perceive at a physiological level much of the fluctuation that goes on in the external world.

Conversely, organisms have techniques of reacting to the rates of change of the external world rather than the actual levels of resources. Water fleas are sometimes sexual and sometimes asexual. They change from nonsexual reproduction to sexual reproduction when a drastic change occurs in the environment, say, the change in the amount of oxygen in the water in which they live or a change in its temperature or a change in food availability. They do not alter from nonsexual to sexual when the temperature is high or when it is low but when it changes rapidly in either direction. They are detectors of change pure and simple. Our visual system is also a sensitive detector of change. Our central nervous system, by complex processing of images, enables us to see differences in intensity of light across edges in a way that is superior to what physical and electronic devices can do. We accomplish this by magnifying differences across small distances. Thus, we have greater visual acuity than optical scanning machinery. The third rule of organism and environment, then, is that fluctuations in the world matter only as organisms transform them.

Finally, organisms actually change the basic physical nature of signals that come to them from the external world. As the temperature in a room rises, my liver detects that change, not as a rise in temperature, but as a change in the concentration of sugar in my blood and the concentration of certain hormones. What begins as a change in the rate of a vibration of air molecules—a change in temperature—becomes converted inside the body into a change in the concentration of certain chemical substances. The nature of that conversion is a consequence of the action of genes, which have a strong influence on anatomy and physiology. When I am out in the desert doing my field work and I hear and see a rattlesnake, those rarefactions of the air that impinge on my eardrums and those photons of light that come into my eye are changed by my central nervous system into a chemical signal and suddenly my adrenaline starts to flow. But these vibrations and photons would be changed to a very different chemical signal in the body of another snake that is receiving exactly the same sights and sounds, especially if it were a snake of the opposite sex. This difference in transformation of one signal into another is coded in the difference between human genes and the genes of a snake. The last rule of the relation between organism and environment is that the very physical nature of the environment as it is relevant to organisms is determined by the organisms themselves.

It may be objected that such an interactive picture of organism and environment is all very well but it ignores some obvious aspects of the external world over which organisms have no control. A human being may have *discovered* the law of gravitation, but he certainly did not *pass* it. You cannot fight gravity. But that, in fact, is not true. A bacterium living in liquid does not feel gravity because it is so small and its buoyant properties free it from what is essentially a very weak force. But the size of a bacterium is a consequence of its genes, and so it is the genetic difference between us and bacteria that determines whether the force of gravitation is relevant to us.

On the other hand, bacteria feel a universal physical force that we do not, the force of Brownian motion. Precisely because bacteria are so small, they are battered from one side to the other by the motion of molecules in the liquid in which they are suspended. We, fortunately, are not constantly reeling from one side of the room to the other under the influence of that bombardment because we are so large. All forces of nature depend for their influence on size, distance, and time duration. How large an organism is, how rapidly it alters its state and position, how far it is from other organisms of different sizes and kinds are all deeply influenced by the organisms' genes. So, in a very important sense, the physical forces of the world, insofar as they are relevant to living beings, are encoded in those beings' genes. Just as we cannot talk about living organisms as just products of their genes but must recognize that the genes interact with the environment in producing the organism in its development and activity, so reciprocally we cannot make the mistake of saying that organisms confront an autonomous external world. The environment influences organisms only through interaction with their genes. The internal and the external are inextricably bound up with each other.

The facts of the relationship between organism and environment have important consequences for current political and social movements. There is a widespread perception that in many ways the world is becoming a rather less pleasant and more threatening place to live in, and there is a good possibility that it may grow catastrophically unpleasant in the not too distant future. It may get a lot warmer. A good deal more ultraviolet light may strike us than now does. The world does not smell very good. There are all sorts of noxious substances that are the agents of illness and even death, and we recognize all these changes as the consequence of human activity. It is entirely correct that human beings should want to make a world in which they can live happy, healthful, and reasonably long lives. But we cannot do that under the banner of "Save the Environment," because this slogan assumes that there is *an* environment that has been created by nature and that we in our foolishness are destroying. It assumes, too, that there is such a thing as the balance of nature, that everything is in a balance and harmony that is being destroyed only by the foolishness and greed of human beings.

There is nothing in our knowledge of the world to suggest there is any particular balance or harmony. The physical and biological worlds since the beginning of the earth have been in a constant state of flux and change, much of which has been far more drastic than anyone can now conceive. Indeed, much of what we conceive of as the environment has been the creation of living organisms. The atmosphere that we all breathe and that we hope we can continue to breathe is about 18 percent oxygen and a fraction of a percent of carbon dioxide. But that atmosphere was not on earth before living organisms. Most of the oxygen was bound up in chemicals. Oxygen is a very unstable compound and does not exist stably in free form. There was, however, a high concentration of free carbon dioxide. The carbon dioxide was removed from the atmosphere and deposited in limestone and chalk by the action of algae and bacteria during the early history of the earth and in oil and coal by plants somewhat later. The oxygen, which was not present at all, was put into the atmosphere by the activity of plants, and then animals evolved in a world made for them by the earlier organisms. Only 60,000 years ago, Canada was completely under ice, as was the middle of the United States. *The* environment has never existed and there has never been balance or harmony. Fully 99.999 percent of all species that have ever existed are already extinct, and in the end all will become extinct. Indeed, life is about half over. Our estimates are that the first living organisms appeared on earth in the order of 3 to 4 billion years ago, and we know from stellar evolution that our sun will expand and burn up the earth in another 3 to 4 billion years, putting an end to everything.

So any rational environmental movement must abandon the romantic and totally unfounded ideological commitment to a harmonious and balanced world in which the environment is preserved and turn its attention to the real question, which is, how do people want to live and how are they to arrange that they live that way? Human beings do have a unique property not shared by other organisms. It is not the destructive property but the property that they can plan the changes that will occur in the world. They cannot stop the world from changing, but they may be able with appropriate social organization to divert those changes in a more beneficial direction, and so, perhaps, even postpone their own extinction for a few hundred thousand years.

Is it within the biological capability of human beings to reorganize their futures? This question brings us back to the issue of human nature and its biological determination. If sociobiologists are right, then human beings have limitations coded in their genes that make them individually entrepreneurial, selfish, aggressive, xenophobic, family oriented, driven toward dominance, self-interested in a way that precludes any real possibility for a radical reorganization of society. You cannot fight human nature. On the other hand, if Kropotkin was right that human beings are biologically impelled toward cooperation and have been artificially held away from it historically, then such a reorganization might be possible. So it would seem that we would need to know the truth about individual human biological limitations. After all, we cannot transcend the limitations that are part of our biological nature. Perhaps we really had better sequence the entire human DNA because that is a first step, although an insufficient one, to learn what human limitations may be. In his book *Sociobiology*, Professor Wilson says,

> If the decision is taken to mold cultures to fit the requirements of the ecological steady state, some behaviors can be altered experientially without emotional damage or loss in creativity. Others cannot ... We do not know how many of the most valued qualities are linked genetically to the more obsolete destructive ones. Cooperativeness towards group mates might be coupled with aggressivity towards strangers, creativeness with the desire to own and dominate. If the planned society, the creation of which seems inevitable in the coming century, were deliberately to steer its members past those stresses and conflicts that once gave the destructive phenotypes their Darwinian edge, the other phenotypes might dwindle with them. In this, the ultimate genetic sense, social control would rob man of his humanity.[1]

It appears, then, that we need to know the genetic linkages between the various aspects of an individual's behavior, because if we do not, we may ruin the world altogether in our blundering attempts to make it better.

The demand for biological information and the implied assumption that society needs to be guided, in the end, by a technocratic elite who understand genetics totally confounds the properties and limitations of individuals with the properties and limitations of the social institutions that they create. It is the ultimate political manifestation of the belief that individual autonomous units determine the properties of the collectivities in which they assemble.

But when we look around at society, we see that the opposite is true. If we have to characterize social organization and its consequences, it is that social organization does not reflect the limitations of individual biological beings but is their *negation*. No individual human being can fly by flapping his or her arms and legs. That is indeed a biological limitation having to do with our size and the size of our appendages. Nor could human beings fly if a very large number of them assembled in one place and all flapped their arms and legs simultaneously. Yet I did fly to Toronto last year, and the ability to fly was a consequence of social action. Airplanes and airports are products of educational institutions, scientific discoveries, the organization of money, the production of petroleum and its refining, metallurgy, the training of pilots, the actions of government in creating air traffic control systems, all of which are social products. These social products have come together to make it possible for us as individuals to fly.

It is important to note that although flight is a social product, it is not society that flies. Society cannot fly. Individuals fly. But they fly as a consequence of social organization.

Sherlock Holmes once explained to Dr. Watson that he did not know whether the sun went around the earth or the earth went around the sun because it made no difference whatsoever to his affairs. He analogized the mind to a kind of attic in which one could put just a certain amount of lumber, and every new fact added had to displace an old one. There is indeed a limit to what any human being can remember if by "remember" we mean the number of things that one can pull out of one's head. No historian of health and disease can remember all the bills of mortality, all the demographic statistics since the nineteenth century. Yet historians do remember those facts because they can look them up in books, and books are a social product, as are the libraries that hold them. So social activity makes it possible for us to remember what no human being could remember as an isolated entity.

Individual biological limitations understood from viewing individuals as isolated entities in a vacuum are not individual limitations for individuals embedded in society. It is not that the whole is more than the sum of its parts. It is that the properties of the parts cannot be understood except in their context in the whole. Parts do not have individual properties in some isolated sense, but only in the context in which they are found. The theory of human nature that searches for that nature in the products of genes in individuals and the limitations of individuals caused by those genes, or in the properties of an external world that are fixed and that cannot be altered except in a destructive way, misses the whole point.

It is indeed the case that human social and political organization is a reflection of our biological being, for, after all, we are material biological objects developing under the influence of the interaction of our genes with the external world. It is certainly not the case that our biology is irrelevant to social organization. The question is, what part of our biology is relevant? If one were to choose a simple biological property of human beings that was of supreme importance, it would be our size. The fact that we are somewhere between five and six feet tall has made all of human life possible as we know it. Gulliver's Lilliputians, who were said to be six inches tall, could not, in fact, have had the civilization that he ascribed to them because six-inch-tall human beings, no matter how they were shaped and formed, could not have created the rudiments of a technological civilization. For example, they could not have smelted iron. They could not have mined minerals, because a six-inch-tall being could not get sufficient kinetic energy from swinging a tiny pickax to break rocks. That is why when babies fall they do not hurt themselves. Nor could the Lilliputians have controlled fire, because the tiny twigs that they could bring to a fire would burn up instantly. Nor is it likely that they could have thought about mining or have been able to speak, because their brains would be physically too small. It probably takes a central nervous system of a certain size to have enough connections and enough complexity of topology for speech. Ants may be terribly strong and terribly clever for their size, but their size alone guarantees that they will never write books about people.

The most important fact about human genes is that they help to make us as big as we are and to have a central nervous system with as many connections as it has. However, there are not enough genes to determine the detailed shape and structure of that nervous system nor of the consciousness that is an aspect of that structure. Yet it is consciousness that creates our environment, its history and the direction of its future. This then provides us with a correct understanding of the relation between our genes and the shape of our lives.

Our DNA is a powerful influence on our anatomies and physiologies. In particular, it makes possible the complex brain that characterizes human beings. But having made that brain possible, the genes have made possible human nature, a social nature whose limitations and possible shapes we do not know except insofar as we know what human consciousness has already made possible. In Simone de Beauvoir's clever but deep apothegm, a human being is *"l'être dont l'être est de n'être pas,"* the being whose essence is in not having an essence.

History far transcends any narrow limitations that are claimed for either the power of genes or the power of the environment to circumscribe us. Like the House of Lords that destroyed its own power to limit the political development of Britain in the successive Reform Acts to which it assented, so the genes, in making possible the development of human consciousness, have surrendered their power both to determine the individual and its environment. They have been replaced by an entirely new level of causation, that of social interaction with its own laws and its own nature that can be understood and explored only through that unique form of experience, social action.

Notes

1/ A REASONABLE SKEPTICISM

1. C.B. MacPherson, *The Political Theory of Possessive Individualism* (New York: Oxford University Press, 1962).

2/ ALL IN THE GENES?

1. R.J. Herrnstein, *I.Q. in the Meritocracy* (Boston: Atlantic–Little, Brown, 1973), 221.

2. L.F. Ward, "Education" (manuscript, Special Collection Division, Brown University, Providence, RI, 1873).

3. A.R. Jensen, "How much can we boost I.Q. and scholastic achievement?", *Harvard Educational Review* 39 (1969): 15.

4. C.C. Brigham, *A Study of American Intelligence* (Princeton: Princeton University Press, 1923), 209–210.

5. H.L. Garrett, *Breeding Down* (Richmond, VA: Patrick Henry Press, no date).

6. H.F. Osborne, letter, *New York Times*, 8 April 1924, 18.

7. R.C. Lewontin, S. Rose, and L.J. Kamin, *Not in Our Genes* (New York: Pantheon, 1984), 101–106.

8. L.J. Kamin, *The Science and Politics of I.Q.* (Potomac, MD: Erlbaum, 1974).

9. Ibid.

10. Jensen, *op. cit.*

11. B. Tizard. "IQ and Race," *Nature* 247 (1974): 316.

12. R.C. Lewontin, *Human Diversity* (San Francisco: Scientific American Books, 1982).

3/ CAUSES AND THEIR EFFECTS

1. E.M. East and D.F. Jones, *Inbreeding and Outbreeding* (Philadelphia: Lippincott, 1919).

4/ THE DREAM OF THE HUMANE GENOME

1. Committee on Mapping and Sequencing the Human Genome, *Mapping and Sequencing the Human Genome* (Washington, D.C.: National Academy Press, 1988).

   Daniel J. Kevles and Leroy Hood (eds.), *The Code of Codes: Scientific and Social Issues in the Human Genome Project* (Cambridge: Harvard University Press, 1992).

   Jerry E. Bishop and Michael Waldholz, *Genome: The Story of the Most Astonishing Scientific Adventure of Our Time—The Attempts to Map All the Genes in the Human Body* (New York: Simon and Schuster, 1990).

   Lois Wingerson, *Mapping Our Genes: The Genome Project and the Future of Medicine* (New York: Dutton, 1990).

   Joel Davis, *Mapping the Code: The Human Genome Project and the Choices of Modern Science* (New York: Wiley, 1991).

   Christopher Wills, *Exons, Introns, and Talking Genes: The Science Behind the Human Genome Project* (New York: Basic Books, 1991).

   Dorothy Nelkin and Laurence Tancredi, *Dangerous Diagnostics: The Social Power of Biological Information* (New York: Basic Books, 1991).

   David Suzuki and Peter Knudtson, *Genethics: The Ethics of Engineering Life* (Cambridge: Harvard University Press, 1990).

   Daniel J. Kevles, *In the Name of Eugenics: Genetics and the Uses of Human Heredity* (Berkeley: University of California Press, 1986).

   The New York Times, 9 April 1992, p. A26; *The Wall Street Journal*, 17 April 1992, p. 1; *Nature*, 9 April 1992, p. 463.

2. Remarks made at the First Human Genome Conference in October 1989. Quoted by Keller in "Nature, Nurture, and the Human Genome Project," in *The Code of Codes*.

3. Pressure against the paper was also brought by scientists in the genome sequencing establishment on the editor of the journal in which it was to be published, including one of the contributors to *The Code of Codes*. As a result, the editor delayed its publication, demanded changes in galley proofs, and asked two defenders of the method to write a counter-attack. One report of the scandal is given in Lesley Roberts's "Fight Erupts over DNA Fingerprinting," *Science*, 20 December 1991, pp. 1,721–1,723.

4. Based on *Frye* v. *United States* 293 F. 2nd DC Circuit 1013, 104 (1923).

5. Committee on DNA Technology in Forensic Science, *DNA Technology in Forensic Science* (Washington, D.C.: National Academy Press, 1989). The reader should know I am not a disinterested party either with respect to the report or to the body that sponsored it. I have twice testified in federal court on the weaknesses of DNA profiles, am the author of a position paper that was a basis for the original very critical version of the NRC report's chapter on population considerations, and am the author, with Daniel Hartl, of a highly critical paper in *Science* that was the object of considerable controversy. I resigned from the National Academy of Sciences in 1971 in protest against the secret military research carried out by its operating arm, the National Research Council.

6. See M. Allison, "The Radioactive Elixir," *Harvard* magazine, January–February 1992, pp. 734–75.

5/ A STORY IN TEXTBOOKS

1. E.O. Wilson, *Sociobiology: The New Synthesis* (Cambridge, MA: Harvard University Press, 1975).

2. Ibid., 562.

3. Ibid., 561.

4. Ibid., 119.

5. Ibid., 556.

6. Ibid., 575.

7. Ibid., 552.

8. Ibid., 554.

9. P.A. Kropotkin, *Mutual Aid* (1901), Chapter 1.

10. E.O. Wilson, "Human Decency Is Animal," *New York Times Magazine*, 12 October 1975, 38–50.

11. *Exploring Human Nature* (Cambridge, MA: Education Development Center, 1973).

6/ SCIENCE AS SOCIAL ACTION

1. E.O. Wilson, *Sociobiology*, 575.